THE UNCERTAINTY EFFECT

Michelle D. Lazarus

MONASH
UNIVERSITY
PUBLISHING

Published by Monash University Publishing
Matheson Library Annexe
40 Exhibition Walk
Monash University
Clayton, Victoria 3800, Australia
publishing.monash.edu

Monash University Publishing: the discussion starts here

9781922633415 (paperback)
9781922633422 (pdf)
9781922633439 (epub)

A catalogue record for this
book is available from the
National Library of Australia

Text design and typesetting by Akiko Chan
Author photograph by Gerard Hynes
Illustrations on p. 25 and p. 144 by Rhys Davies
Images on p. 41 and p. 82 reproduced with kind permission
from The Noun Project
Diagrams on p. 5, p. 208 and p. 232 by Kat Orgallo, Zenzi Design
Population dynamics diagram on p. 121 reproduced with permission
from Geoff Boeing. It originally appears in his 2016 paper 'Visual analysis
of nonlinear dynamical systems: chaos, fractals, self-similarity and the
limits of predictions', published in the journal *Systems*.

Printed in Australia by Griffin Press

CONTENTS

Introduction   *1*

Chapter 1: Questions in the Classroom   *9*
How teachers can foster uncertainty tolerance in students

Chapter 2: Treating the Unknown   *47*
How those in healthcare can manage uncertainty

Chapter 3: Counting on Uncertainty   *96*
How those in economics and business can deal with the unknown

Chapter 4: Unravelling Life's Mysteries Collaboratively   *131*
How uncertainty affects scientific discovery

Chapter 5: Cultural Fluency   *179*
How uncertainty tolerance can foster diversity and social inclusion

Chapter 6: The Age of Uncertainty   *237*
How uncertainty tolerance helps us navigate local and global challenges

Acknowledgements   *269*
Bibliography   *273*

INTRODUCTION

'Living with uncertainty is one of the few established facts
of modern life.' —LAWRENCE J. RAPHAEL

When you think about the last time you experienced uncertainty – when you were faced something unknown or unclear – how did you react? What did you think? How did you feel? What did you do? The answer to these questions is what defines your level of uncertainty tolerance.

Uncertainty tolerance starts with a 'stimulus', or a source of uncertainty. This stimulus can be a situation with incomplete, inconsistent or conflicting information, or a situation so utterly complex it becomes challenging to identify a clear way through.

Now is a useful time to write this book because we all experienced a shared uncertainty stimulus in recent times when the Covid-19 global pandemic hit. So much was unknown at the beginning. Who was at risk? How did the virus spread? Were there any effective treatments? Who should we listen to – politicians, scientists, self-styled 'Covid experts' touting cures ranging from light exposure to drinking bleach?

As the virus spread and lockdowns lengthened, the complexities and conflicting information only increased. People across the world wondered: how accurate is lockdown modelling? Is there a scientific basis for 'herd immunity'? How do we prevent viral spread to developing countries and ensure equal access to vaccines across cultures and continents? This left many of us unsure how to proceed in our daily lives. We had to weigh a series of expert opinions – and not all the experts were saying the same thing – alongside inconsistent media headlines and our neighbours' views.

The way that individuals responded to the pandemic in its early stages tells us a lot about uncertainty tolerance. At the start of 2020, nearly everyone was seeking to manage their anxiety. Some did so by claiming that Covid-19 was 'mild' or 'just like the flu', while others sounded alarms about its lethality and refused to leave their homes. Experts debated the mode of transmission (droplet versus airborne) and how useful masks were in preventing the spread of infection. With each mutation of the virus, there was a predictable conversation about whether this new strain was more transmissible and lethal than the previous. At times it seemed that new variables came by the hour. We all had massive uncertainty in our personal and professional lives. What varied was how we responded to this: our level of uncertainty tolerance.

Uncertainty tolerance, or our responses to the wide-ranging stimuli of uncertainty that Covid-19 generated, varied across populations and cultures. Some of us responded with curiosity, reviewing the data as more research was shared and knowledge was developed. This cognitive-based response might have been accompanied by a feeling of confidence in our ability to handle the uncertainty that the pandemic presented, despite our inability to answer every question we had about the virus. Together, this set of responses represented a high level of tolerance of the pandemic's sources of uncertainty. Others responded with anxiety and fear, and either became paralysed by inaction or denial, or went into a flurry of seemingly illogical activity, including panic-buying. These behavioural responses to the pandemic represent the other end of the uncertainty tolerance spectrum – an *in*tolerance of uncertainty.

I experienced a tempered intolerance of uncertainty. I purchased toilet paper, but just enough for us (and our neighbour) to last a couple of months. I am a scientist by training, so I was less anxious about the changeable nature of the science. I knew that with such a novel virus, conflicting information was to be expected. Despite this, I was now watching the nightly news with anxiety. Every day a new curveball was thrown. On a Monday we were told to stay home if we could, wash all food we brought into the house and practise good hygiene; a week later, we were in hard lockdown and had a nightly curfew. The changeable nature of the government's response to the pandemic became a source of concern – one that countered the other factors that made me more tolerant of the uncertainty.

When I later asked a range of Australian science and medical academics how they responded at the start of the pandemic, most said they felt a variation of the Australian phrase 'she'll be right'. This made me smile. It represented the ethos of a culture where people went about their business as usual, trusting that everything would be fine eventually. This response was no doubt helped by a feeling of safety due to early border closures, low case numbers and the geographical benefits that being an island continent provides.

Why was my response so different from my colleagues'? In part, it related to my personal circumstances during the pandemic.

I am an expat from the United States living and working in Australia. I had family in the age group considered vulnerable in locations where the pandemic was raging uncontrolled. These factors, known as **moderators**, shifted my uncertainty tolerance towards the negative end of the spectrum. Would my anxiety have been lessened if my personal circumstances had been different – if my family were all based in the same country as me, where they were less at risk of infection? Possibly. This raises the question of how much our tolerance of uncertainty comes from 'nature' (is built into our personalities) and how much is 'nurture' (derived from our circumstances and cultural context).

Many researchers in the past have suggested that we each have an innate level of uncertainty tolerance, meaning we naturally gravitate towards being more tolerant of uncertainty (that is, curious, confident, able to make decisions despite the uncertainty) or less tolerant (that is, anxious, vulnerable and paralysed or in denial in the face of uncertainty). But contemporary research suggests that our uncertainty tolerance is an interplay of nature and nurture. We may be born with an innate tendency towards one end of the uncertainty tolerance spectrum (nature), but the unique circumstances in which we find ourselves when faced with an uncertain stimulus (nurture) also play a role.

For instance, consider my response to the uncertain stimulus of the pandemic. When it comes to my family's health, I tend to be naturally more anxious when uncertainty is present, representing a less tolerant position on the uncertainty tolerance spectrum. This baseline is tempered by my experiences with the intrinsic uncertainties in scientific research, as I am keenly aware that science changes as new information comes to light, and this slides my uncertainty tolerance towards the positive end of the spectrum.

My family being overseas tips the scale back towards the lower end of the spectrum. In combination, I ended up being a little more towards the negative end in my response to the uncertainty present in the early days of the pandemic. However, if my uncertainty tolerance was solely based on my natural, 'innate' responses to uncertainty, I would likely have been squarely categorised as *intolerant* of uncertainty. This illustrates how combined factors impact our cognitive, emotional and behavioural responses to uncertainty to generate our overall uncertainty tolerance. Moderators work in combination to position each of us on the spectrum of uncertainty tolerance.

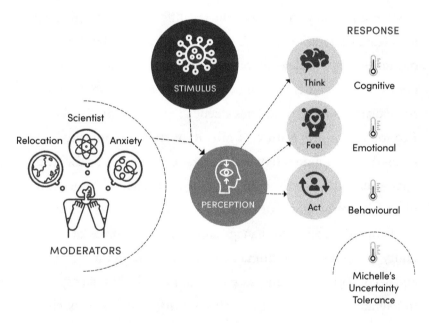

The uncertainty tolerance conceptual model shows that uncertainty tolerance starts with the perception of an uncertain stimulus – an event or a situation that is ambiguous or complex. The individual responds by adapting how they think, feel and behave. The individual's perception and responses to the uncertain stimulus are influenced by factors such as personality, the situational context and sociocultural practices. Contemporary research suggests that our level of uncertainty tolerance isn't static, but changeable over time and across contexts.

But uncertainty, and our ability to tolerate it, isn't only relevant in times of great stress, such as pandemics. Uncertainty is part of daily life and work, as the pithy quote from communications sciences professor Lawrence J. Raphael illustrates. One of the greatest sources of uncertainty is humans – the unpredictability of our thoughts, emotions and behaviours. No matter how many times researchers look to history to predict the future, or model data to determine outcomes, the nature of human autonomy, agency and choice generates universal uncertainties.

Humans both generate uncertainty and must learn to manage the uncertainty we create. As our lives become more global, and as we interact daily with those who are different from us (through social media, traditional media or travel), knowing how to manage our uncertainty or develop our uncertainty tolerance becomes a critical life skill. It is also an important proficiency for workers, particularly those in industries steeped in technology. The World Economic Forum's Future of Jobs Report, issued in 2020, estimates that 85 million jobs will be lost to technology worldwide by 2025 – and that's just the start. Existing artificial intelligence (AI) often fails to recognise or respond to uncertainty. At this point in history, the ability to manage sophisticated sources of uncertainty is still unique to humans. As AI becomes more advanced and becomes increasingly present in our lives, the human capacity to tolerate and even thrive with uncertainty becomes an essential workplace skill that protects and futureproofs our jobs.

This book explores how we can better manage uncertainty in our lives, our workplaces and our relationships. Drawing on the latest research, it shows how shared sources of uncertainty manifest in diverse ways across different industries and fields, from education to medicine to the business and technology worlds.

While these workplaces may appear to have little in common, there are shared features in the way workers across all of them cope with uncertainty. For instance, a lower tolerance of uncertainty tends to lead to employee burnout, regardless of industry (or job description). Uncertainty intolerance is also linked to a more rigid workforce that struggles with flexibility and going 'off-script'. At its worst, uncertainty intolerance can create a culture of workers who entirely lack the skills to be flexible when the circumstances call for it – as they do more than ever in a fast-paced, changeable world. This can have ripple effects across an economy and a society, such as high staff turnover rates and/or exclusion of workers who don't 'conform' to the rigidity set by their less uncertainty-tolerant colleagues. But if we can seek to better understand and embrace uncertainty, the benefits are many. These pages explore how fostering and supporting a greater tolerance of uncertainty in ourselves and our communities can help us to create more inclusive and culturally sensitive societies, which are potentially more adaptable to the troubling existential questions facing us, such as planetary health, global economic crises and even future pandemics.

Case studies are embedded in each chapter to illustrate the ways in which uncertainty manifests in different fields and how we can learn to navigate it. I include evidence-based recommendations on how to prepare current learners for future workplace uncertainties. In exploring the basis of uncertainty tolerance, factors that influence it, the impacts on our personal and professional lives, and practical ways we can support ourselves (and others) to develop uncertainty tolerance, this book serves to support our capacity to not only survive uncertainty but to thrive in it.

The research in this book is drawn from the broader literature on uncertainty tolerance as well from as my team's own research. Over the last five years, my team, comprising diverse researchers and PhD students from fields such as business and economics, healthcare, science and psychology, have been working together to collect data from learners, academics and workers to better understand how uncertainty manifests in workplaces and how we can better prepare the future workforce for these uncertainties. I have applied many of the lessons learned through this work to my own teaching, and as a leader in support of staff as they journey through uncertainty. The positive impacts I saw from translating these research findings into real-world, everyday contexts compelled me to write this book.

As the Indian spiritual guru Sadhguru stated, 'Those who try to avoid uncertainty only end up avoiding possibilities.' My goal in this book is to help us all identify when we are facing uncertainty and to think about how to embrace both the uncertainty and the certainty to increase our tolerance for change, for personal wellbeing and the greater good. *The Uncertainty Effect* is a spark that seeks to ignite a broader conversation about how to improve our workplaces, our social interactions and our daily lives against the constant background hum of uncertainty. It can also shine a light on how we move past uncertainty and inaction when it comes to the key global challenges confronting us as a species, from climate change to pandemics.

So let's find out together what we can learn from uncertainty tolerance.

1

QUESTIONS IN THE CLASSROOM

How teachers can foster uncertainty
tolerance in students

*'Education: the path from cocky ignorance to
miserable uncertainty.'* —MARK TWAIN

The year 2020 seemed to start as usual. I'd just returned from an annual medical education conference in Singapore and was deep in planning for Semester 1. While a few attendees at the conference had mentioned a 'pandemic' they had heard about in the news, I'd paid little attention. The media love to hype viruses and bugs, so I figured this was just fear-mongering. I had heard this type of headline before, with swine flu and bird flu. My focus was on getting course manuals printed and ensuring everyone knew which lectures they were giving, and where they needed to be on campus, come the first day of class.

Yet by orientation week, just a month later, everything had changed. The novel coronavirus had hit Australian shores and taken hold along the east coast of the mainland. Governments were in the early response stages; decisions and actions were changing by the day, even by the hour. No one seemed to know what was going on. Panic was setting in among the community. A campus-wide bulletin was sent out: semester was being delayed a week.

I was being peppered with questions. Would we be teaching in person? If so, would there be safety measures in place? If not, could we still give lectures on campus? Would they be live-streamed? If teaching online, how would we run practical classes? What technology was available? Was there funding for new equipment? What were the students being told to expect? The questions kept coming, and I struggled to provide answers.

It came down to the wire. The weekend before the delayed start of semester, the messaging was still at risk of change. Communication was fractured across the entire higher education sector. Our institution was, like others, struggling with the enormity of the uncertainty. My staff and I didn't know if we should be teaching to more than 600 students per class face-to-face with a novel virus in circulation, or if we should instead move all classes online or even cancel them entirely. As teachers of medicine, nursing and health sciences, we had some insight into the virus and were worried. I got a call on Sunday evening from a staff member, concerned for her health and that of her young children. During that call, I made a decision and told the staff member I would take the heat. We were moving online.

By now I had read the data and looked at what was happening in other parts of the world. Despite the suffocating uncertainty, I knew a captain's call was necessary. I had to put into practice what I taught and what I researched; I had to become more tolerant of uncertainty, of making decisions when there was no way to know *all* the relevant information or whether a decision was the *right* one.

For all of us teaching at that time, from primary to secondary to tertiary education, the pandemic brought a tornado of uncertainty. There were technical problems related to the mode of teaching and access to resources. There was fear over the sheer complexity of hastily migrating decades-long face-to-face teaching approaches into online formats. There were questions about the learner's environment: would students have what they needed to learn? How would they receive support if stuck? Would they engage properly with their learning?

The pandemic also brought about a greater awareness of the challenges of education because more of the community was now doing the teaching. Teachers were no longer the sole deliverers of content; now parents and siblings were responsible for helping children to learn. Many were realising just how uncertain education can be.

Before the pandemic, most people probably pictured education as a teacher communicating knowledge to an eager (or at least present) group of students. This teacher might be in front of the class writing on a whiteboard, or beside a screen presenting digital slides. The mental images we conjure in these scenarios typically represent a 'sage on the stage' didactic, teacher-centred modes of learning, where content is transmitted from the expert to the student in a way that reinforces knowledge as certain. Even student-centred teaching approaches, where students are actively engaged in their own learning, have certainty built into them. Active learning environments typically rely on activities with predetermined end points, which students work towards solving. Rarely do these activities have a flexible or unknown end point, challenging students to actively learn to manage uncertainty.

In today's classroom, it seems there is little scope for uncertainty. How many students have heard a teacher complain about 'getting through the material' or lament the lack of time to deliver content? Often the focus is on communicating knowledge, not on preparing learners for uncertainty. During the pandemic, many teachers prioritised delivering information over skill development. We were trying to concentrate on what we thought was paramount: certainty. But what if, throughout our schooling, we had been prepared to manage uncertainty? Would we have handled the pandemic with greater ease and less discomfort?

Today's students will face a world brimming with uncertainty, even without a pandemic. It will affect their jobs and their daily lives. We can help foster learners who are better prepared to face future social and workplace challenges by realigning our traditional views of teaching, and by incorporating evidence-based opportunities for learners to practise coping with uncertainty in low-stakes environments (where their careers and others' lives and livelihoods are not at stake). I refer to this as uncertainty tolerance pedagogy, and the aim is to encourage students to feel a level of comfort with facing and managing uncertainty.

Why teach uncertainty tolerance?

Over a decade ago I was sitting outside my first clinical anatomy exam, feeling overwhelmed. I knew I had gaps (more like huge, gaping craters) in my knowledge of anatomy. I was filled with anxiety about everything I didn't know. All the thoughts running through my mind were negative: *Maybe I'm not cut out for this; maybe I've been out of university for too long. Is it too late to quit?*

I was completing a postdoctoral fellowship in anatomy and it had not been easy. I was in a new state, a new culture, a new environment. It was my first time below the Mason–Dixon line, and others told me they could tell I was a 'Yank' 'just by the way I walked'. I had realised during my studies that real anatomy is quite different from any textbook illustration. Moreover, anatomy can differ from person to person, so I was uncertain both about the sources I should be using to study and about their content. It had been years since I was in a formal classroom. I had been wrestling with feeling deeply out of practice, unsure how to even go about studying again, and struggling to figure out what was relevant to learn in such a broad (and detailed) area. Being new, I couldn't even figure out who to ask for help.

There I was sitting on the floor outside the anatomy lab, in hospital scrubs, fifteen minutes before exam start time. I was panicking, flipping between pages of handwritten scrawls in my notebook over and over, futilely trying to soak up every last fact in the few moments before I had to walk inside and identify anatomical structures on real people.

Understanding a learner's experience can help guide teachers in considering how, when and why to integrate uncertainty into our teaching. My experiences in anatomy education are echoed by participants in every field my team studied. Learners face diverse and numerous sources of uncertainty. These uncertainty stimuli, as I call them, can include uncertainty about the learning environment (particularly if unfamiliar), about what is relevant to study and about who to turn to for help. Until this anatomy exam, I had a sound system for learning reinforced by good grades. But being in a new, unfamiliar context had me questioning my abilities. I had moved from familiarity to discomfort – from certainty to uncertainty.

My team's research and the broader literature on uncertainty reinforces my experiences in studying anatomy. Many learners encountering uncertainty, particularly for the first time, are distressed. They are likely, at least initially, to feel discontented, resentful, annoyed and sometimes even angry. But just because something is challenging doesn't mean we shouldn't teach it. Our job is to help prepare students for the world they will face beyond the classroom. This includes learning to cope with uncertainty. We can do this by offering students opportunities to recognise uncertainty and then teaching them how to navigate it, ultimately helping them to become more adaptable.

There are many benefits to developing uncertainty tolerance in students. The qualities teachers report seeing include: humility, as students begin to realise their views aren't the only ones possible; cognitive flexibility, whereby they adjust their understanding with new information; and confidence in their ability to manage future sources of uncertainty. While students may not always enjoy the journey, most report being grateful for the opportunity to practise uncertainty tolerance and feel more capable of handling future encounters with uncertainty in their studies or in life.

Many teachers are already fostering uncertainty tolerance in the classroom. Some ideas in this chapter, particularly in the next few pages, will be familiar to the engaged teacher. Yet this chapter helps us to think about the goals and purposes of guiding students towards managing uncertainty, and to look at how teachers can build an entire educational program focused on helping learners develop their uncertainty tolerance. It also offers those who are new to the idea of uncertainty tolerance some practical and accessible ways to foster it in the classroom.

Building a curriculum that fosters uncertainty tolerance

Uncertainty tolerance is related to what renowned higher education professors Jan Meyer and Ray Land call a 'threshold concept'. Coined in 2003, the term refers to knowledge or learning that forces the learner out of their comfort zone (away from familiar ways of thinking and doing) and into discomfort (towards new ways of thinking and doing). Central to this is the change a learner experiences after working through the uneasiness that uncertainty brings. They will no longer see the world the same way; they are themselves different once they have crossed the 'threshold'. And, as more adept thinkers and learners, they will be better prepared to confront complex, changeable situations.

The Covid-19 pandemic is an excellent example. Those of us who lived through it have moved from what was 'normal' in 2019 to a 'new normal' in 2023, signifying new ways of thinking and doing in this mid- to post-pandemic world. These changes are both big and small. Before Covid (BC), few except germaphobes paused to think about the hygiene implications of blowing out candles on a birthday cake. Most of us happily sang aloud in close quarters and blew our moist, droplet-filled saliva all over a cake to be shared with others. Now, this practice seems repulsive to many – or at least gives us pause. Our views of this ritual, as well as our behaviours, have changed. BC, we might have greeted even a relative stranger with a hug, a handshake or a kiss – now there is often awkwardness, a wave or a verbal acknowledgement. As a society, we have crossed a threshold. We are not the people we were in 2019 and will never be again. We can wish for life to be more like it was in 2019, but we can't go back. This transformation is irreversible. But it was essential as we grappled with the emergence of a dangerous virus and shifted our thinking to accommodate it.

An analogy I sometimes use to consider threshold concepts in education involves a tunnel. When we introduce students to uncertainty in a classroom context, it is as though we are guiding them from a sunny, clearly marked trail to the mouth of a dark tunnel. Some will cross the threshold eagerly, curious about what they might encounter beyond; others will be more cautious, and some will simply refuse to enter.

If tunnels bring thoughts of claustrophobia, consider instead a swimming class. The instructor is teaching a group of children to swim for the first time. If he throws them all in the deep end, some might learn to float, but a lot are going to panic and sink. Of those who sink, some may actually start to drown. Those who survive may be so scared that they never get back into the water again, ruining their chance to discover how much they could enjoy it.

In either analogy, the teacher can play an important role in the way students learn to navigate uncertainty. There are two steps in fostering uncertainty tolerance: introducing uncertainty and showing students how to navigate it. Both stages are vital to ensure that students can experience uncertainty but are not left feeling overwhelmed by it. An educator I interviewed summed this up well: 'I know that uncertainty gives me power, doesn't it? Because now I know I can choose to include [uncertainty tolerance teaching practices] as I want.' In the tunnel analogy, the teacher could choose to spend time preparing students to enter the tunnel and then focus on lighting the way as they traverse it. In the swimming class, the instructor could teach some basic floating techniques, maybe on dry land, so that the children feel more confident about entering the pool. The goal of uncertainty tolerance pedagogy is to create an environment that helps students cross the threshold from certainty to uncertainty without getting lost or drowning. If the process is done right, students will develop the problem-solving skills and creative capacity to find solutions to complex problems. They will build confidence in doing so, meaning that future experiences of uncertainty become less daunting.

How to stimulate meaningful uncertainty in the classroom
Developing learners who can manage uncertainty effectively requires careful planning. The first step is to stimulate uncertainty – teachers must bring students to the threshold of the tunnel. Each uncertainty stimulus helps students begin to learn that knowledge – how it is understood and used by individuals and society – has an inherent degree of uncertainty.

Each teacher can decide when in the learning journey is best to introduce uncertainty. In every classroom there are many pre-existing sources of uncertainty beyond the educator's control (such as learners' prior knowledge, socio-economic conditions and so on). Some classes may be ready to tackle uncertainty from the outset; others will require a little more time to sit with certainty before entering the territory of the uncertain. Even within a single class there may be students more ready than others to face uncertainty. Regardless of this diversity, my research team has identified some simple teaching strategies educators can use to stimulate uncertainty in their learners.

1. **Question learner preconceptions**. Teachers can challenge learners to move towards uncertainty by asking questions about their fundamental beliefs and truths. The focus here is on challenging long-held ideas central to how the learner sees and understands

the world. For instance, one teacher described a simple exercise that involved bringing a chair to the front of the room and commenting that our society has defined the chair as an object for sitting on. But we could just as easily define this chair structure as something for standing on, or for storing books on, and so on.

If, for example, you are teaching a science class, try asking students to define terms like 'living organism'. What makes something alive? How do we know it is living? This approach generates uncertainty by stimulating learners to think about the assumptions they hold. Students may have taken for granted that everyone defines the terms 'living' and 'alive' the same way. A discussion with the whole class, which embraces multiple points of view and different ideas and theories, allows students to learn from each other that there isn't one simple, universal definition to this question – or to knowledge more generally.

2. **Draw on the experiential learning spectrum.**
    A type of teaching first defined in the humanities and social sciences, and typically referred to as 'cultural literacy pedagogy', is centred around experiential learning – where students engage actively in their own learning through experiences.

The term 'cultural literacy' was first coined in 1988 by E.D. Hirsch, a professor of education and humanities, who described teaching approaches that help students learn to understand the traditions, behaviours and history of a cultural group, including one's own. Cultural literacy's focus on the 'human factor' provided a nomenclature for what my research team found about stimulating uncertainty across all disciplines, from science and mathematics to healthcare to the humanities to business – people are the principal source of uncertainty stimuli.

Dr Gabriel Garcia Ochoa, a senior lecturer and leader in cultural literacy teaching practices at Monash University, along with Professor Sarah McDonald, Academic Director of Student Experience at Monash University, helped to refine cultural literacy in the classroom into three phases. In each phase, a different number of uncertainty stimuli is present.

- *Critical incident*: so termed because of the potentially 'critical' challenge a student encounters, critical incidents have the least amount of uncertainty stimuli of all three phases. If we return to the tunnel analogy, a student experiencing a critical incident will encounter the dark passageway but can see the light at the end – so the tunnel is dimly lit instead of pitch black.

In classrooms, critical incidents are usually presented in the form of case studies or examples. Students are exposed to a limited degree of uncertainty related to the unknowns of the case. The case could be designed to have unknowable components that students need to grapple with, or a twist where something about the case changes as the students work through it. Often, there are multiple 'correct' or reasonable answers. Learners are tested to challenge their assumptions, but only so far.

Another way to build critical incidents into learning is through using raw data or real-world situations instead of constructed examples and case studies in textbooks. This way, the case already has built-in 'messiness' and uncertainty, but also a resolution in the form of the course of action taken, which can be debated and discussed – keeping with the idea of the dimly lit tunnel.

• **Destabilisation**: this includes activities in which the sources of uncertainty increase to a point where a student can begin to feel destabilised. Instead of just reading about a case that has uncertainty stimuli and working through the case-derived uncertainties, students take this a step further and act out a case or scenario. The uncertainty comes from both the ambiguities and unknowns in the case and from the interactions and decisions of the individuals in the role-playing team. There is now added uncertainty about how teammates will act as they work through the case together to solve it.

Role-playing exercises or simulations can introduce destabilis-ation. These activities cannot be solved simply by following a series of steps, as in a recipe, but are designed to include unknowns. They are often complex, requiring independent thought and creative problem-solving, and may give rise to several possible solutions. For example, physics students could be tasked with building a bridge that supports x weight with y parameter/restric-tion (length, width or building materials). This activity requires students to understand and apply basic principles of physics, but also allows opportunities to practise dealing with uncertainty. Here, one uncertainty stimulus is the multiple ways students can construct the bridge. Another is how the group works together to build their chosen design.

At the end of the activity, the bridges could get tested and each group share their approach to the task and its challenges, illus-trating the variable paths a group could take to find a solution. To further stimulate uncertainty, the teacher could ask each team to describe what they would change if the bridge needed to support more weight. In this way, the teacher is stimulating uncertainty first with destabilisation (working in a team to solve a problem physically) and then with a critical incident (working as a team to solve a problem theoretically).

- **Iso-immersion**: in this phase of experiential learning, there are more sources of uncertainty than we can count. Think of 'work-integrated learning' or 'work-based placements', such as classroom placements for trainee teachers or study abroad for students. In teacher placements, for example, the uncertainty stimuli can include the students' understanding of the trainee teacher's role, the novelty of the environment and uncertainties about how to teach the content or who to go to with questions. The trainee has a lived experience of the innate uncertainties associated with a workplace – no matter how much this teacher-in-training knows, the way in which they will apply their knowledge is highly variable, and depends on the classroom context in which they are placed. In our analogies, iso-immersion is pushing learners straight into the darkness of the tunnel or throwing children in the deep end of the pool. For this reason, iso-immersion is often best introduced to learners who have had prior opportunities to cope with managing uncertainty.

Importantly, a teacher doesn't need to go through these experiential learning phases sequentially or according to a certain timetable. Depending on who the learners are and what the teacher is trying to achieve, they may decide that iso-immersion on Day 1 is appropriate. Or they may start students in destabilisation, move them into iso-immersion, and back to critical incidents. The key is to keep learners questioning and curious – and away from the complacency of certainty, which is only ever an illusion in our uncertain world.

3. **Introduce a range of perspectives**. When learners realise that knowledge can be 'in the eye of the beholder', they begin to understand the limits of certainty. By exploring a single topic from multiple points of view, learners start to realise that events can be interpreted in various ways, and therefore any knowledge derived from experience and interpretation carries with it a certain level of uncertainty.

An example I use in educator workshops is a diamond. Not only are these cut to be multifaceted to the eye, but we can consider them in a variety of ways. At first glance, a diamond seems fairly 'certain' – it is a physical object with particular characteristics. But if we assembled a panel of individuals to discuss diamonds, this notion of certainty starts to fade. For instance, a scientist might describe a diamond in terms of its carbon make-up; someone who just got engaged may speak about what a diamond signifies in their culture; someone who lived in a war zone may describe diamonds in terms of their impact on conflict. And so on. These views aren't just different sides of a coin; the significance and purpose of the object itself differs in each consideration. This panel, with their range of perspectives, would illustrate the uncertainty intrinsic in people's perceptions and interpretations of knowledge.

Teachers can introduce multifaceted perspectives by inviting a panel of speakers, rather than a single 'expert', to the classroom to discuss a given topic. Members of this panel could all have a perspective of value – if the topic was food production, for instance, members might include a farmer, a consumer, a supermarket worker and a researcher. The teacher can facilitate the panellists to speak both to the commonalities of their views and to the points of difference. Students can then debate whose perspective on the topic is most valuable, and/or how to reconcile the different perspectives, to develop a presentation that synthesises the diversity of views.

The next step is encouraging students to always think of contrasting viewpoints on a topic, or alternatives to their own modes of interpretation, to ensure they are not being blinded by their own desire to cling to certainty.

How to moderate uncertainty and move beyond paralysis

Once students understand that knowledge is not always certain, the next stage of teaching uncertainty tolerance focuses on helping them to find their way. Teachers can guide students through the murkiness that uncertainty gives rise to with 'moderating' teaching practices. This second stage is crucial. If teachers simply stimulate uncertainty without moderating it, students will be left without the tools to progress in their thinking. By choosing which moderators to use and when to use them, teachers can help students move away from doubt and paralysis, and instead learn to embrace and adaptively respond to uncertainty.

Learners resistant to uncertainty may feel anxious about taking *any* step across the threshold into the darkness of the tunnel and may do everything they can to stay on the familiar trail they are used to walking. They may wish (sometimes vocally) that the class could be taught differently, or that the content could give rise to clear answers. These hesitant and resistant learners are the key focus of moderating teaching practices. My team's research has identified that moderating teaching practices can have a powerful impact on such learners' ability and willingness to work towards tolerating uncertainty.

In workshops with educators, I sometimes illustrate the role of moderators by showing a picture of a dark, cave-like tunnel and asking participants to describe how they feel when they imagine being in this environment. Responses span a spectrum, from 'curious', 'excited' and 'hopeful' to 'scared', 'cautious' and 'anxious'. When I share a picture of a tunnel with some lit areas and some dark areas, participants almost uniformly respond that they feel excited – often filled with awe, as they can see the scope and scale of what was previously hidden and are curious to see what else can be illuminated. By simply shining some light into the darkness, a teacher can foster in students a desire and motivation to explore.

We can think of moderators as sources of light in the dark tunnel. Some learners who have experience with tunnel or cave exploration will light up their own torches to help navigate the darkness. Other students will be novice spelunkers and will rely on the teacher's torch, at least until they acquire the confidence and skills to light the pathway themselves.

When building learning environments that foster uncertainty tolerance, educators can select which sources of light to turn on, which to turn off, and at which point the lights should shine the brightest. Different moderators are needed at different times for different students – some may become bored if the light shines too brightly; others might be terrified in the dark. Making thoughtful decisions about what, when and why moderators are used is where the teacher's expertise comes into play. If the teacher also understands what sources of light learners hold, they can build this into their planning, perhaps grouping students with different levels of experience in managing uncertainty in ways that allow them to support one another. Much like leading a group hike in real life, where each person carries different materials, an educator building uncertainty tolerance pedagogy into their teaching needs to be aware of who has access to which moderators.

Teachers stimulate uncertainty by bringing students across the threshold into a dark tunnel. Once students are inside, teachers can select which moderators (lights) to turn on to help them illuminate a path through.

Students most prepared to traverse unfamiliar terrain are those with prior knowledge and experience – high subject mastery – and/or who are intrinsically motivated to learn. The students who stand at the tunnel's edge with a limited ability to light their own way, and who are more likely to pause or resist walking through the darkness, tend to have low subject mastery and be merit minded. Novices, in either the content or the learning environment, will be hesitant to navigate uncertainty. My experiences in my postdoctoral fellowship illustrate how novelty influences the perception of uncertainty. I was in a new city and a new culture. I was researching a field foreign to me as part of a brand-new lab group, working with team members I didn't know well. I remember telling my husband, 'Every day is just an opportunity for me to illustrate how much I *don't* know.' I was lost, having entered a dark tunnel illuminated by only a few shards of light, and the sheer depth and breadth of the novelty disoriented me. If at this point an educator had come along and purposefully stimulated more uncertainty – instead of turning on some lights – I don't think I would have made it through. I think I would have quit: turned around and tried to find my way back to the familiar ground I knew.

Merit-minded learners, meanwhile, tend to focus on external motivators for learning, such as grades or awards. While some teachers may instinctively view this characteristic as undesirable, the reality is that for many learners, assessment is high stakes. They believe that grades and awards will determine their entire future. If they don't get the desired grade, they fear this will limit their career options or result in losing privileges (even love) at home. Whether their belief is correct or not, for these learners, grades carry weight and carry risk. Merit-minded learners tend to be hesitant because unfamiliar, uncertain surroundings mean putting at risk highly effective, familiar study techniques that produce expected outcomes. These learners will need more light to navigate the tunnel. My team's findings suggest that reassurance and pastoral care are the other most effective means of support here. This pattern among learners is not only present in the classroom: in subsequent chapters, we will see that perceptions of risk play a significant role in uncertainty tolerance in workplaces too.

The clever and compassionate teacher needs to consider the learner's contextual uncertainties. Students with low subject mastery are already in the darkness, and if they are exposed to large amounts of uncertainty through iso-immersion, for instance, they will only be plunged deeper into an abyss. Similarly, if it is the first day of class and students are grappling with the demands of a new environment, educators may choose to delay purposefully stimulating more uncertainty.

Introducing moderators into the classroom

My research team has identified more than twenty teaching practices that educators can employ to help students learn to manage stimulated uncertainty. Here are some that almost any teacher can introduce into the classroom.

- **Orientate students to reduce the mental load.**
  A theme that kept cropping up across all the data my team collected was the importance of reducing extraneous sources of uncertainty to allow learners to focus on the relevant educational uncertainty. Teachers can give an orientation to the learning environment as a simple way to reduce unnecessary uncertainty. For instance, on the first day of class, spending extra time explaining how the classroom works, who to go to for information, where the toilets are and so on lessens the level of uncertainty students are navigating due to the novelty of the unfamiliar environment. This frees students to embrace the purposeful uncertainty the teacher stimulates in the classroom.

- **Be transparent about the likelihood of discomfort.**
  Help prepare learners for the uncertainty they will experience in class by being clear that moving away from the familiar and expected – away from certainty – is not easy. To quote one teacher: 'You have to set expectations that these [learning experiences] WILL be uncomfortable, WILL feel very … it could feel painful.'

Teachers who acknowledge that uncertainty comes with discomfort better equip their students to accept this discomfort when it appears. Conversely, teachers who see their students struggling with uncertainty and tell them to 'get on with it', or show frustration when a student disengages, generate uncertainty intolerance in their learners. In this context, students begin to become intolerant of uncertainty in the classroom because they think they are unable to deal with it 'correctly' and so seek to avoid engaging with uncertainty entirely.

- **Help learners develop a sense of purpose.** As students flounder in the darkness, teachers can support students to focus on *why* they need to navigate the tunnel. If they have a purpose for travelling into the unknown and a reason for their discomfort, learners are more likely to take that first step and to continue. Teachers can point to the fundamental nature of uncertainty across different careers and explain how learning to manage uncertainty now will help prepare students for the workplace. If entering the workforce is too far removed from the students' learning journey, teachers can instead point out that learning to manage uncertainty in this activity will help students complete their next activity more easily. One teacher I interviewed described how she stimulated uncertainty and afterwards discussed it with her students: 'Why are we doing this? Why is this relevant to your future?' Here, the teacher isn't actually focusing on why the knowledge is important to their future, but rather on why the struggle with uncertainty is. As another teacher put it: 'It's really critical that we do raise some of these things in class with our students; it's just a sea of ambiguity [otherwise], and it's going to make them feel really anxious. And ignoring it, buying into that defence system, is unhelpful and irresponsible on our part as teachers.'

Without motivation, students may lack the drive to continue grappling with uncertainty. My own intolerance to uncertainty at the start of the pandemic was alleviated once I focused on my sense of purpose. As a leader, I found a reason to manage my uncertainty in supporting the staff through *their* uncertainty. Knowing I had to manage my uncertainty to assist others was enough to foster my own uncertainty tolerance.

- **Foster diverse teamwork.** Instead of asking students to tackle problems individually, or in groups of like-minded students, teachers can divide the class into diverse teams. My team's research suggests that teamwork alone, without building diversity into the teams, does not have the same impact on learners' uncertainty tolerance as diverse teamwork. Teachers should aim to generate teams of students from different backgrounds, cultures, levels of socio-economic status, language groups and so on. The goal of these diverse teams is to reduce the echo-chamber effect, which produces unilateral solutions to problems. This approach also accounts for pre-existing variability in students' ability to deal with uncertainty.

A team with diverse perspectives can help each member recognise that there are many ways to approach a problem, re-inforcing that there is uncertainty in how we perceive, understand and interact with the world. An important first step is for the team members to become familiar with one another. Teachers can devote some class time to having students learn about one another's backgrounds. Students will begin to realise that having access to different knowledge bases and a variety of experiences can help them to navigate uncertainty with more confidence. Teachers can reinforce this by highlighting how the team's diverse experiences and ways of thinking can be drawn on to address sources of uncertainty. By illustrating that 'we are stronger together' when navigating uncertainty, the teacher helps set a precedent for learners engaging in diverse teamwork later in their careers.

- **Outline clear roles for each team member.** Another way to reduce unnecessary uncertainty is to task team members with clearly defined roles. The extent of the guidance provided on these roles depends on the students' skill level. Are the learners novices? If so, the teacher might need to detail what different roles entail – for instance, a note-taker compared to a communicator. If the students are more experienced learners, they may only require an instruction to identify and define the roles needed to complete the task – because these students can handle the uncertainty of the task as well as the uncertainty of how to manage the task. If a task has large amounts of uncertainty (i.e. iso-immersion), providing clear roles for the learners can help them better manage the intrinsic uncertainties of the complex environment.

- **Create spaces conducive to navigating uncertainty.** Not surprisingly, being singled out or called upon in front of the class, where an answer will be heard and judged by others, tends to lower students' tolerance of uncertainty. In contrast, anonymity – for instance, anonymous discussion boards – can enhance students' tolerance of uncertainty because it creates a space to experiment and to be 'wrong' without judgement. Learners in an anonymous learning environment tend to embrace uncertainty to a greater extent because the risk related to this uncertainty is lower.

- **Support learners' wellbeing.** Uncertainty is challenging for everyone, especially when the stakes are high. Teachers who acknowledge this and offer empathy and support, providing learners with a sense of hope, can help alleviate the emotional toll that dealing with uncertainty can take. Teachers can share their own similar experiences of uncertainty, indulging in 'intellectual candour', to help the learner understand that they are not alone. They can also explain the strategies they use to navigate uncertainty. This support can help students feel grounded and empowered. Importantly, the discomfort associated with uncertainty is not synonymous with an unsafe learning environment, and supporting learners' wellbeing is critical.

- **Engage in reflective practice.** A key method of data collection in my team's research was reflective diaries and interviews. These provided learners with opportunities to describe how they thought, felt and behaved when faced with learning situations they perceived as 'certain' and 'uncertain'. Time and again, participants highlighted how contemplating their experiences helped them improve their tolerance of uncertainty. In their reflective diaries, students used strong adjectives to describe the moments they experienced uncertainty in the classroom. They described feeling frustrated, overwhelmed, disheartened, anxious or even guilty. Yet as they reflected on the process through their writing, and later in interviews, these students frequently highlighted their gratitude in learning skills to manage uncertainty, and acknowledged their growth in coping with uncertainty even when describing the same situations they had previously perceived as negative.

Importantly, our research revealed that reflective practice about uncertainty tolerance is most effective when it is not graded. Teachers may choose to incorporate reflective practice into their teaching either by only assessing these reflections for completion or by separating them from the coursework entirely.

Weaving moderators into learning

Moderators of uncertainty tolerance are not all-or-none – research suggest they are additive, each building on the effect of the other. Moderators can be combined and used selectively at different points in the learning journey to help students manage uncertainty. For instance, with novice students who are already experiencing a lot of uncertainty, adding in opportunities to anonymously play with knowledge can be useful. For more advanced students, teachers may increase the responsibility for knowledge – these students may be tasked with presenting in front of the class or called upon to provide their perspective during a debate. Reflective practice can enhance uncertainty tolerance in all groups and skill levels, even when students are novice learners, and can be integrated across the entire learning journey.

Let's see how moderators can be employed across a series of classes. Imagine it's the first week of the school year and students are in a science class learning about cells. After an around-the-class introduction and a comprehensive orientation from the teacher, she begins to discuss the idea that science is not certain. She talks about the discomfort that dealing with uncertainty will at times bring nearly everyone in the class, coupled with a clear explanation of why she has chosen a style of teaching that emphasises uncertainty – because it will foster a tolerance of uncertainty, and so help students to prepare for future challenges, inside and outside the science classroom. In this, the teacher gives a sense of purpose to the learning.

In the next class, the teacher leads students to the tunnel's threshold by stimulating uncertainty. The students are learning about organelles – the organs of a cell – and their functions. This textbook knowledge provides the learners with some certainty. The teacher then tasks the class to design a new, undiscovered cell that has one key improvement on cells as we know them – for instance, it can divide at a faster rate, or it differentiates atypically. This critical incident will stimulate uncertainty by forcing students to move from what is 'known' to exploring what is unknown (and may be unknowable). By this point the tunnel is already dimly lit, even before students enter, because the teacher has turned on the 'orientation', 'sense of purpose' and 'discomfort transparency' lighting.

The teacher decides to turn on a few more lights, because this classroom is filled with novice learners. Instead of having students work alone or in pairs with the person sitting next to them, she divides the class into diverse teams. Here students will together cross the threshold of the tunnel when designing their novel cell. The teacher also assigns roles to each member of the group: researcher, note-taker, illustrator and communicator. The communicator has the greatest level of responsibility for knowledge. This is a chance for the teacher to see how those students called upon to be communicators handle the extra pressure and uncertainty of sharing the 'unknown' answers (given that this is a novel cell).

In the third class, learners work in teams to design their cell, and the communicator presents their results to the class. Unsurprisingly, each team designs a radically different cell. At the end of the activity, the teacher asks the teams to describe how and why they made the decisions they did to develop their ideal cell, and closes by highlighting how the different approaches are all (at least in part) effective solutions. To further stimulate uncertainty, the teacher asks each team to swap their cell design with another team and tasks the other team with improving on the design. This is another critical incident. Points are awarded for the most innovative result to encourage motivation and teamwork. At the end of the class, one team is declared the 'winner' of the challenge.

In the fourth class, the teacher holds a whole-class discussion as a form of reflective practice, asking groups to articulate the challenges of the classroom task. This illustrates the role of uncertainty at the same time as reinforcing their knowledge of the functions of a cell's organelle. The teacher also takes this opportunity to highlight elements of a cell's organelles that are still unknown or debated in the field.

This content could have instead been delivered as a series of lectures. Yet didactic teaching tends to inhibit learners' uncertainty tolerance by reinforcing knowledge as finite and knowable. Instead, balancing factual presentations of knowledge with relevant uncertainties helps students to learn that uncertainty isn't distinct from knowledge but rather inherent in it. This concept is extended in how teachers facilitate knowledge.

My team's research suggests that when fostering uncertainty tolerance in novice students, it is important to address questions in a way that highlights areas of contention, or 'grey areas', instead of just supplying a simple answer – as this didactic approach reinforces knowledge as static facts, too. When educators explain how answer x is sometimes correct, and other times answer y is correct, students become more comfortable with the inherent uncertainties of knowledge. The teacher in our example really 'shined' in accomplishing this.

How to assess uncertainty tolerance

The way in which teachers choose to assess learners can affect a learner's tolerance for uncertainty – particularly for those who are merit-minded. Modes of assessment reveal what is valued in the classroom. We cannot expect to develop uncertainty-tolerant learners if all assessments are rigorously scripted, with no representation of ambiguity or unknowns. Learners will not be able to illustrate their ability to deviate from the norm or to show creativity and initiative – or to have high-stakes opportunities to practise managing uncertainty.

Teachers I spoke to often struggled to identify a direct way of measuring learners' development of uncertainty tolerance through traditional classroom-assessment strategies. Many used proxy assessment tasks that evaluated learners' capacity for communicating the intrinsic uncertainty in applied knowledge. These might be discursive assessment tasks, such as asking students to highlight what is known, debated and unknown in a field, or challenging them to pick a side of a heated debate, substantiating their argument with existing knowledge while recognising its limitations. Others develop assessments with future-focused or fantasy problems, where learners are asked to apply existing knowledge to offer solutions to a problem that doesn't yet exist – a bit like the example I provided earlier with the novel cell.

One way uncertainty can be built into assessment is to allow students to choose the form of their assignment. For example, learners could be tasked with describing different cell organelles (certainty), but allowed to choose how they present this (uncertainty). The presentation could take the form of a poster, a podcast, a talk, a video recording or a song.

Essentially, teachers who are successfully fostering uncertainty tolerance in their students are using open pedagogical approaches with less prescriptive guidelines, flexible assessments and/or eliminating grading entirely. The absence of prescriptive guidelines is key, as overly detailed instructions can cause students to become intolerant of uncertainty when there is a discrepancy in instructions, because the rest seems so 'certain'.

When guidelines are left open, students may initially question everything, but once they learn that they have the right to take initiative, they often embrace the uncertainty. As one teacher explained:

> There's been a very big push for our assessment tasks to be very, very clear and instructional, pretty much stepped out stage-by-stage. I took that away from my students this year. I gave them an assignment where I said, 'You have five options (a podcast, a poster, online news article, video – whatever) and I don't care what format it comes in.' I decided I was going to trust them, and I said, 'What you have to do is learn to live with the uncertainty of not having instructions.' And I have to say they completely overwhelmed me with their ability to just embrace it. All of my colleagues said, 'This is never going to work … This is going to fail miserably because you don't have enough instructions for the students' … We [the course teachers and I] just reassured the students at almost every point, 'Don't be afraid to be experimental. Don't be afraid to be creative. If you have any questions, ask us.'

This teacher's comments highlight that flexible assessment parameters encourage creativity and adaptable thinking, hallmarks of an enhanced capacity to manage uncertainty.

Interestingly, my team found that 'single best answer assessments' – multiple-choice questionnaires with one correct answer, for instance – had variable impacts on learners' uncertainty tolerance. Some students found this style of assessment helped them by providing boundaries around what content was relevant. Others reported that it hindered their ability to focus on uncertainty because it forced them to find a clear answer. The latter point is why many medical schools in the United States and some Australian universities – such as the University of New South Wales and Swinburne University of Technology – are moving towards 'ungrading'.

Ungrading refers to assessments that are marked on a pass or fail grading system. A pass indicates competency in a given field or body of knowledge. At Swinburne, student work is assessed in line with one of three categories – meeting, exceeding or failing to meet expectations – with ongoing feedback about the students' performance given throughout the semester. In many US medical schools, a pass score indicates 'competency for healthcare practice', nothing more or less. While this may seem controversial, since it does not stratify students into ranks based on results, I believe that this is a trend we will soon see more widely because this practice more authentically replicates the real world – where you are benchmarked on your competency within a role, and your performance ideally improves progressively through ongoing experiences and feedback. As a colleague commented, 'As a patient, I care more that the doctor is competent than whether they received high-class honours during their degree. In fact, it wouldn't even occur to me to ask.'

Ungrading supports the developmental process by rewarding growth, as opposed to a final mark. Given that uncertainty tolerance has a large developmental component, ungrading is a valuable way to assess students' progress. A news article showcasing Swinburne's transition to ungrading included a telling student quote: 'Assessment for learning makes me feel more comfortable experimenting with new skills, even when I might be uncertain of the final outcome.' Ungrading does indeed generate the space needed for learners to become more tolerant of uncertainty.

A 2018 study conducted by US researcher Brittany Ange and her team found that ungrading had no negative impacts on academic performance in medical schools – where grades have traditionally been considered very important. Combined with Darcy Reed's 2011 study, which showed that pass–fail grading had a large positive impact on students' psychological wellbeing (greater than even the type of learning approaches used to teach the content), we can see that ungrading may hold value as a moderator in a learner's journey towards uncertainty tolerance. If the goal is to help learners develop uncertainty tolerance, the way we reward and acknowledge this progress is another means to light the path to success.

How teachers experience uncertainty

Teaching is filled with intrinsic uncertainty. To quote one of our research participants: 'In my experience, there is no one classroom; there is no one approach to every classroom. Classrooms, from day to day, from lesson to lesson, will vary depending on so many things – things visible and invisible; things that are the result of interactions which may have occurred in the yard, at home, or even things which occurred in the student's past that you may be unaware of. So, when I think about uncertainty, I think that's almost like the normal situation of a classroom.'

This sentiment is echoed by Rosemary Hipkins, chief researcher at the New Zealand Council for Educational Research. She highlights that any number of factors could impact on what the student is learning, regardless of the teacher's intended goals. How can an educator be expected to plan effectively for every circumstance, every context and every learner's needs? Can any educator truly know all the moderators a learner brings into the educational environment? At any given moment teachers are both experiencing uncertainty and needing to manage and support their students' experiences of uncertainty.

Preparing students for managing uncertainty inevitably introduces uncertainties for the teacher. There is (of course!) a degree of uncertainty in developing a curriculum that fosters uncertainty tolerance. While this can be daunting, especially for those who have typically developed their teaching activities in reference solely to syllabus guidelines, it also offers a sense of freedom to employ those same qualities we are seeking to foster in students: creativity, initiative, risk-taking.

What we are talking about here is a refresh. Educators need to develop their own tolerance of uncertainty to survive, let alone thrive, in their jobs, just as much as their students need to develop uncertainty tolerance to future-proof their forthcoming careers. If teachers are struggling with managing their own uncertainty, they will also struggle with the uncertainty central to the type of teaching we are discussing, and to teaching more generally. Teachers need to feel empowered and confident to develop the learning practices that best suit their specific classroom context. So how to do this?

One way to navigate the uncertainty embedded in teaching, US researchers Michelle Jordan, Rober Klensasser and Mary F. Roe suggest, is engaging in reflective practice. My team's research revealed similar results: educators' critical reflective practice (reflecting on their teaching and exploring how current experiences may impact on their future actions) can help them develop, and purposefully integrate, uncertainty tolerance teaching practices into their curriculum more effectively. Time and again, teachers I spoke to who were fostering uncertainty tolerance in the classroom highlighted the need to first be tolerant of uncertainty themselves. Again, we see how reflective practice fosters uncertainty tolerance both in the classroom and in the workplace.

In my research, I asked teachers to reflect on the different ways they stimulate uncertainty in the classroom, and how they support learners as they move through this uncertainty. I queried them on how their students responded to this type of teaching. One teacher observed that until he reflected on his uncertainty tolerance teaching practices through this research, he hadn't considered the pivotal role these practices play for learners: 'It actually made me reflect much more on it than I've done before – because I have to think about why students may struggle with it.' Others echoed the impact of reflective practice in helping them manage the uncertainty of their students' responses to the teaching. One said, 'It's really … made me think a lot about what I do and why I do it. And what seems to work and why sometimes things don't work. And if I think about when I was … first starting, [it] would have been really helpful to have some of these concepts front of mind, because it's kind of what I came up against … at least now I understand. It gives me a better framework for [understanding] why these students were reacting the way they were, which is useful … I know the reason behind why they are so distressed.' And another: 'It's certainly made me start to think about it in a more practical way. I think I use some of these principles in my teaching, but I've never done it particularly consciously.'

Many teachers felt a desire to build more activities designed to foster uncertainty tolerance into their teaching, through inquiry-based learning, complex problem-solving and other active, experiential learning approaches. The challenge was how to do this with purpose.

BENEFITS OF REFLECTIVE PRACTICE FOR EDUCATORS

My team's research revealed that reflective practice enhanced teachers' management of uncertainty by:

- providing a shared language
- increasing awareness that introducing uncertainty into the classroom had real purpose
- reinforcing their capacity to integrate uncertainty tolerance pedagogy into curriculum design
- reducing their own uncertainty, giving them a renewed sense of purpose.

How to value and support teachers

It's all very well to encourage teachers to embrace uncertainty. But it's a near impossible task if our workplaces aren't set up to facilitate this. Educational systems and institutions need to support teachers to do their work well by fostering their continued development.

My team's research revealed that, over and over again, teachers came up against systemic barriers when trying to develop a curriculum that foregrounded uncertainty. Sometimes this was peer nay-sayers. Other times it was the overly prescriptive guidelines, particularly around assessments, set by educational bodies. Teachers described how benchmarking assessments, which are meant to provide a standardised view of learning, prohibited them from adjusting lesson plans to incorporate uncertainty tolerance teaching practices. The tiered grading system, which rewards certainty, was another challenge. Educational systems aim to lessen the embedded uncertainties in education by generating frameworks and competencies, and having educators engage in tick-box lesson plans. These attempts at conformity often generate frustration and cause burnout for those in the classroom – because no matter how hard a teacher works to reduce uncertainty, education is a human activity and uncertainty is part of the human condition.

The fact is, we need a fundamental shift in educational systems to support uncertainty tolerance teaching practices, as well as more teachers engaging in such practices. Modern education systems need to reduce the expectation of certainty in a world where there is none. Institutions should work towards reducing the frequency of, and value placed on, standardised examinations and benchmarking. Instead, they could allow for some flexibility in reporting learning outcomes, and relax overly prescriptive guide-lines that provide a false sense of certainty and result in lowering uncertainty tolerance in learners and educators.

An example of this is student evaluations of teaching (SET). Higher education institutions typically engage SET to provide structured feedback from students to teachers and use them to benchmark learning and quality of teaching. However, every student's educational journey is different, particularly when the education is structured around uncertainty. Some educators reported moving away from teaching practices that foster uncer-tainty tolerance to those that hinder it because of negative SET comments. Time and again, teachers find that students experi-encing discomfort with uncertainty tend to see the curriculum and the teaching style as the source of their pain and unhappi-ness: 'I've had a lot of SET responses where they're [the students] saying, "Just give me a straight answer." It's been throughout my career ... SETs coming back with, "Don't be ambiguous. Give me a straight answer." I've had that often. And that clearly shows that there are some students who don't understand what it's all about.' And: 'There's quite a lot of backlash from students in the SET. So the [teacher] didn't perform very well in the SET. Which, I think that's a barrier. Well, [uncertainty and ambiguity] was built into the curriculum.'

There is often a fundamental disconnect between teachers and students, fostered by their institutions. Educators are having to battle against the institutional guidelines to create an environ-ment that helps build uncertainty tolerance in learners. Given this style of teaching challenges students, SET feedback may be more related to the discomfort accompanying the challenge than the impact of the learning in preparing them for the future.

Even the timing of SET can make a difference. Mid-year, when students are still steeped in uncertainty, is when most are likely to be experiencing discomfort and anxiety. By the end of semester, students are more likely to feel positively about their learning experience. A 2018 study exploring the timing of teaching evaluations found more favourable views of teaching at the end of a term. SET may be best timed at the end of a teaching year, or even after the students have completed the class.

We also need to help novice teachers focus on their purpose in teaching uncertainty tolerance. Whether it's considering the school motto, the impact on learner development or the results of workforce participation upon graduation, it is key for educational leaders to connect what is done in the classroom to the wider philosophy of creating capable learners. Transparency about the fact that education is rife with challenging uncertainties can also help teachers to feel supported. Toxic positivity – where educational leaders (principals, chancellors, school presidents) – portray the uncertainty teachers experience as exclusively 'exciting' and an 'opportunity for growth' may be counterproductive, especially for novice teachers. Various studies have shown that teachers who struggle with managing embedded uncertainty are more likely to burn out. My team's data suggests that one cause of burnout may be the teacher's attempts to bridge the disconnect between the educational system's focus on uniformity and standardisation and the real-world practice of teaching, where uncertainty results from heterogeneity and diversity. In many of the interviews I conducted, I could hear the frustration teachers feel when facing a system of conformity against a backdrop of intrinsic uncertainty.

In the transition to remote learning during the pandemic, leaders who acknowledged the intrinsic uncertainty of teaching in unprecedented circumstances (and the related discomfort) – instead of repeating 'Everything is fine!' – had higher levels of teacher engagement. Furthermore, teachers who were given the most freedom (that is, the least prescriptive guidelines) had more capacity to address the uncertainty related to teaching in a pandemic. Essentially, these educational leaders were moderating their teachers' uncertainty tolerance towards the positive end of the spectrum. The uncertainty-tolerant teachers were more likely to be creative and curious when seeking solutions in an unknown landscape. The challenge for the future is to not relegate such practices to 'unprecedented times', as everyday teaching is filled with uncertainty.

If we don't move towards a more flexible, less prescriptive education system – one that can support teachers' capacity to foster uncertainty tolerance – we will continue to put strain on an already stretched workforce, and we will be failing to prepare students for the uncertainties in their futures. There is already evidence of this occurring, with media reports highlighting that secondary-school teachers are leaving en masse due to burnout and a lack of flexibility among their employers, and reports of students who feel underprepared for the workforce.

Light at the end of the tunnel

Preparing both learners and teachers to cope with the uncertainties implicit in education is essential for wellbeing and career preparedness. My team's research, combined with the broader body of literature, suggests that we have the agency and power to do this through the way we teach as individuals and the structures embedded in education systems.

While some of the teaching practices in this chapter may seem intuitive, my team's research illustrates that what enhances or inhibits uncertainty tolerance the most is whether teachers have the agency and the ability to *purposefully select* when, where and how these teaching practices are employed, both within individual activities and across the learning journey. The goal for teachers is to plan and execute uncertainty tolerance teaching within a system that fosters their own tolerance of uncertainty.

Building uncertainty tolerance takes pacing and attention. Over time, students develop a permanent skill set to manage uncertainty. While few of us enjoy facing uncertainty, particularly when the stakes are high, with practice we are more likely to become grateful for the skills we develop in navigating it, humble in recognising the limits of our own knowledge and more capable of tackling the next uncertain experience. To quote one teacher: 'It's about humility in understanding your own strengths and weaknesses.'

By fostering uncertainty tolerance in our learners and teachers, our society will be more equipped to 'go where there is no path and leave a trail', as philosopher Ralph Waldo Emerson put it. In a fast-changing world, when our future depends on our ability to navigate through and thrive in unprecedented times, this is essential. Instead of treading our familiar route, due north to uncertainty may just be the compass point we need to set.

TREATING THE UNKNOWN

How those in healthcare can
manage uncertainty

'Medicine is a science of uncertainty and an art of probability.'
—WILLIAM OSLER

n the classic Agatha Christie story *The Cornish Mystery*, the Belgian detective Hercule Poirot proclaims through his thick moustache: 'A doctor who lacks doubt is not a doctor. He's an executioner.' Leaving aside his assumption that doctors are male, Poirot's statement challenges a widespread assumption. Crime and medical television dramas reinforce the idea that there is only one accurate diagnosis of a sick patient, or a single, incontestable cause of death – which can be discovered in thirty minutes to an hour when just the *right* doctor orders just the *right* test. But this is a fantasy. In real life, online healthcare discussion boards are filled with patients who describe waiting months to years before a diagnosis comes – if one is ever provided – alongside heart-wrenching stories about misdiagnosis leading to potentially devastating long-term consequences.

Back in 2017, my father called to tell me about an article he had read in the local paper. The 'Australian flu', as it was called, was just beginning to enter circulation in the United States. This flu strain was particularly potent and was being blamed for a reported increase in flu-related deaths. While the article was brief, I think the word 'Australian' triggered his desire to share it with me – a link between the United States and Australia helped him to feel connected to his expat daughter.

The story revealed that a mum took her newborn to the doctor. Both had flu-like symptoms: runny noses, congestion and fevers. The doctor's physical exam, which included listening to their breathing, indicated that influenza was likely. Both mother and baby were given a rapid flu test. The tests came back negative, and the doctor thought both were probably infected with another virus in circulation, as is common in the transition between autumn and winter. The doctor suggested lots of fluids and rest, and to return if symptoms worsened. Within forty-eight hours, the article went on, 'both died from the flu'.

I was silent. What a sad case. Then I said, 'This is what I research, Dad. I look at how uncertainty impacts the healthcare profession.'

This doctor had trusted a negative diagnostic test over his physical examination, which revealed clinical signs and patient symptoms that aligned with the flu (and had led him to order the test). The result of believing that the test results were 'certain' was lethal for these two people. The article closed by highlighting how this case resulted in changes to protocols around the use of the rapid flu test, particularly doctor training about the intrinsic uncertainty of the results: these tests aren't 100 per cent accurate (not even close, actually) but the accompanying testing paperwork didn't explicitly highlight this. Much like a rapid Covid-19 test, a rapid flu test can be a useful tool for making a diagnosis in combination with symptoms, but it isn't on its own a definitive source of a diagnosis. Poirot's words rang in my head: when healthcare providers don't leave room for uncertainty, the results can be deadly.

Uncertainties in patient care

A multitude of factors cause uncertainty in healthcare. Patient presentations differ, as symptoms depend on how an illness presents in an individual patient. Unfamiliar diseases or new strains of virus pose additional unanticipated challenges. Take Covid-19: a novel virus that seems to manifest differently in different people. We still don't fully understand the long-term health impacts of infection – this aspect of the pandemic will likely remain shrouded in some degree of ambiguity for decades to come.

Uncertainty doesn't only come from the biophysical nature of illness but also from the interactions between healthcare provider and patient. An encounter between a doctor and a patient is embedded with doubt. How will the patient's previous healthcare experiences impact on their frame of mind during the medical consult? What is the doctor's prior experience with cases of similar presentation? What is the patient hoping to hear? What is the doctor wanting to share? What are the doctor's values, and how might they inform the health advice given and the communication with the patient? How much of what the doctor says will the patient absorb and remember? The variabilities in a single visit to the doctor are too numerous to control.

Then there is the uncertainty of the healthcare system itself, a complex and (in many countries) underfunded leviathan. What are the rules and regulations that a provider and patient must work within? What are the values of the system, and how do they align with the patient care required? How are different specialities set up to interact with each other? The uncertainty of the healthcare system impacts on the uncertainties in patient care, creating meta-uncertainty. For instance, doctors practising in rural communities typically have limited interaction with the hospital system, which can lead to gaps and unknowns in the communication of patient information. Another example: whether or not a patient has insurance may incentivise certain forms of patient care over others.

In fact, the further I got into exploring where and how uncertainty presents itself in healthcare, the more I began to wonder: is there anything about healthcare that *is* certain? At the core of medicine is humans – patients, healthcare providers and caregivers – and the innate uncertainties of how humans will act, think and feel fills the world of healthcare with uncertainty.

Consider a patient who arrives to an Emergency department with chest pain. Ambiguity begins in the first moments – when considering the source of the chest pain. Google 'causes of chest pain' and you will get everything from indigestion, chest infection, rib fracture, blood clot, muscle strain and inflammation to angina, aortic dissection, shingles (chicken pox) and heart attack. Then there are the uncertainties related to the healthcare system. The culture and level of care possible at each hospital is highly variable. Is this a public or private hospital? Is the hospital in a city, or is it a rural community urgent care facility? Each of these factors will determine the capacity of that hospital and what is possible to achieve.

In this case, the triage nurse at a large city public hospital decides to send this person for review immediately because of the potential for life-threatening risk, despite there being many causes of chest pain that are not life-threatening.

Once in an Emergency department bed, the patient undergoes a physical examination, and the healthcare assessment team considers a series of tests in an attempt to narrow down the biophysical uncertainty. What tests should the team order? To order all available tests related to chest pain is impractical and, depending on the patient's level of healthcare coverage, expensive. Some tests can even cause harm, and the team must weigh up the potential benefits versus the risks. In this case, a CT angiogram might be recommended to explore the coronary vessels and blood flow to and from the heart, but such tests involve radiation and there are some people who are allergic to the dye required to visualise the vessels.

The assessment team settles on evaluating heart rate and blood pressure (typical for any patient) and administering an ECG (electrocardiogram) to look at the electrical output of the heart as a proxy for heart rhythm. A nurse comes into take bloodwork to determine whether heart tissue damage has occurred. He also starts reviewing the patient's history for the doctor. The patient is a cis-gendered female, Sally, who is thirty-two years old and exercises regularly.

Sally is whisked away for a chest x-ray. The nurse peppers her with questions, and she struggles to answer and make eye contact through the pain. The doctor, when she arrives, assumes Sally's faltering answers are because she is unsure of her symptoms – and makes the next decision without consulting with Sally because time is critical. Here, uncertainty is introduced due to the communication dynamic between the doctor and the patient. Sally stammers that she has had this pain before but was told she had 'a pulled muscle' and sent home.

The bloodwork comes back and results are within the normal range, but at the high end of the markers for heart tissue damage,

leaving the doctor to interpret whether these findings are significant. Is it because of the timing of the bloodwork that they aren't
higher? Sally doesn't have any of the classic signs of heart attack
that the doctor learned about in medical school (such as crush-
ing chest pain or radiating pain down the arm), but heart attack
can't be ruled out just yet. The standard symptoms and test results
don't seem to apply to this patient. The doctor is confused: is this
a harmless muscle strain or a medical emergency?

Another blood test comes back indicating the potential for a
blood clot, but the doctor recalls that this test often gives a false
positive, suggesting clotting when there isn't any. The ECG shows
some abnormality but isn't conclusive in ruling a heart attack in or
out. A chest x-ray looks clear, which can help rule out other causes
of chest pain, such as pneumonia and rib fractures, but doesn't
exclude clotting as a cause. So now the doctor is left to deal with
conflicting information and ambiguous results.

Without a clear diagnosis, the doctor must decide on a treatment plan. She consults with her peers, including a cardiologist and
a physiotherapist, who offer advice on the phone. All have differing
opinions. She makes the call to keep Sally in hospital to monitor her
and asks the nurse to take additional bloods in a few hours to check
for any changes – any further information could guide a next step.

The doctor wants to defer to an expert and writes up a formal consult with a cardiologist. She has misdiagnosed a patient in
the past who later died from a heart attack. Here, how the doctor
responds to the uncertainties in play is being moderated by her
prior experiences. She refers the patient on and defers the decision on a treatment plan because of this.

Sally is feeling overwhelmed. She is stressed by the uncertainty
of what is causing her chest pain and the sudden news that she will

be staying in hospital overnight for monitoring. The uncertainty is compounded by the lack of information and communication from her treatment team. After a restless sleep, she concludes the next morning that she hadn't communicated well with the doctor the day before. She was in pain and worried she might say the wrong thing. She had been to Emergency before and been told her symptoms were nothing, so she was nervous the doctor would feel she was making it all up. Her prior experiences were colouring her tolerance of uncertainty. When the nurse comes in to check on her later that morning, she is focused on asking to see the doctor, and seems to forget all her relevant symptoms.

Let's leave Sally resting in bed for the moment. Her case highlights the multiple sources of uncertainty, identified in both my team's research and the broader literature, that can be present in every healthcare encounter. Uncertainty stimuli appear in Sally's presentation, the diagnostic findings, the conflicting expert opinions and even in the relationship and communication styles between Sally and her healthcare team. The result in this situation is that the doctor errs on the side of caution, given that Sally's condition could be life-threatening. This case shows how all healthcare providers need to face and manage uncertainty while balancing risk – and their capacity to do this well impacts on patient care and the psychological wellbeing of both the patient and the healthcare team.

One research team leading the way in exploring healthcare-related uncertainty tolerance includes Dr Paul K. J. Han, a senior scientist at the Division of Cancer Control and Population Sciences at the National Cancer Institute; Dr Marij A. Hillen, a specialist in medical psychology and medical communication at Amsterdam University Medical Centre; and Dr Tania Strout, the director of research at Maine Medical Center. This group is changing the way we think about uncertainty tolerance within healthcare systems. Their work led to the first comprehensive model for healthcare-related uncertainty tolerance, presented in Chapter 1. This model moves away from the idea that uncertainty tolerance is something you either have or you don't, and instead suggests that moderating factors – such as the seriousness of the situation – influences individuals' tolerance of uncertainty (for both the provider and the patient).

Sally's case illustrates how aspects of a healthcare situation influence those involved with patient care – Sally had a potentially serious symptom, chest pain, a prior history of Emergency presentations and a doctor who had learnt to be cautious. The hospital setting meant that a variety of tests could be ordered, and peer consultations between specialists were possible. This raises the question: to what extent do different healthcare providers have a natural propensity for being tolerant of uncertainty?

The uncertain healthcare provider

Some research suggests that specialists have varying levels of uncertainty tolerance due to the degree of uncertainty inherent in their field. Psychiatrists, for instance, work in a field of medicine with a lot of ambiguity. Psychiatric diagnosis is typically based on an assessment of an individual's personal history and their current mental state. There are very few 'tests' and biophysical diagnostic tools (like the ones for Covid-19 or flu) for psychiatric illnesses, and this can result in perceptions of a less clear diagnosis and treatment plan. For this reason, many hypothesised that those in the field of psychiatry would have a higher tolerance of uncertainty and that those with a lower tolerance of uncertainty would be more likely to be found in specialities with clearer impacts on patient outcomes, such as surgery. This line of thinking, however, tends to treat the sources of uncertainty that doctors manage across different medical specialties as somehow unique, and different from the uncertainty that other health professions (such as nurses, physiotherapists, radiographers and so on) experience. Ironically, studies exploring the relationship between medical speciality and uncertainty tolerance often yield conflicting results and are themselves uncertain. Some studies show a relationship between uncertainty tolerance and choice of specialty while others do not. Furthermore, no single health profession appears to have a monopoly on uncertainty. From oncologists to podiatrists to occupational therapists to paediatricians, healthcare providers experience an overabundance of uncertainty stimuli. Consider a patient coming to the physiotherapist with a concern about recurring lower back pain. Multiple studies show that lower back pain is a condition imbued with uncertainty: is paralysis possible? Is the cause benign muscular spasms that may go away on their own? Is there an insidious cause, such as cancer? A May 2022 article in *The Medical Journal of Australia* by physiotherapists Emma Ho and Professor Paulo Ferreira claims that 9 out of 10 cases of lower back pain (commonly referred to as 'non-specific low back pain') are not found to be linked to a specific pathoanatomical cause, meaning there is no visible anatomical change that could explain the symptoms. Given that 8.5 out of 10 people will suffer from lower back pain at some point in their lifetime, the uncertainty related to this clinical presentation is commonly encountered. Ho and Ferreira suggest that this uncertainty can 'leave clinicians and patients feeling unclear about the best choice of treatment'. While they do offer some potential solutions, which include a combination of psychological interventions alongside physiotherapy, the day-to-day management of uncertainty with such patient presentations is real and frequent.

In an attempt to reduce uncertainty, those seeing these types of patients most frequently, such as physiotherapists, need to make decisions about patient management with far less power to access diagnostic tests than doctors in a hospital setting. Some studies reveal that physiotherapists manifest their uncertainty tolerance in how they approach patient care, potentially not 'pushing' patients far enough when it is warranted for fear of poor outcomes. This includes pairing back treatment plans to a degree where they are ineffective. Other studies suggest that physiotherapists may have a lower tolerance for uncertainty due to a perception that their patients expect certainty in the form of a clear and definitive diagnosis.

Doctors also experience this. One report indicated that nearly half of primary care patient referrals (including to physiotherapists) are for musculoskeletal pain. Many of these patients will suffer from unexplained pain – where the cause is not identifiable and may never be known – so the pressure for certainty through a diagnosis may not only be inappropriate, it may also be unachievable within our current understanding.

What we can say is that when we consider how uncertain healthcare is, due to the nature of the work, it is reasonable that *all* health professionals who stay in the sector are likely to be at least somewhat tolerant of uncertainty, and also are likely to be faced with frequent and/or substantial uncertainties on a daily basis.

A large-scale 2019 review, led by UK psychiatrist Dr Jason Hancock and leading medical education researcher Karen Mattick, revealed a strong association between an intolerance of uncertainty and psychological distress for healthcare professionals. The inability to manage the intrinsic unpredictability of healthcare may lead to stress, burnout and damage to psychological wellbeing, ultimately leading to negative outcomes for both the patient and the healthcare provider. We'll talk more about this later in the chapter, as burnout among doctors is a crucial issue.

Uncertainty tolerance also impacts on how the healthcare provider interacts with their patients. There is evidence that those with a lower tolerance of uncertainty tend to approach patient care paternalistically. These providers engage in a way that puts them in a position of power and minimises patient autonomy. Doctors are less likely to share their uncertainties about treatment options with their patients and less likely to engage in shared decision-making, where the patient and healthcare provider discuss the options and settle on a plan together. For these healthcare providers, sharing uncertainties introduces new ones to a situation already riddled with unknowns. *Will the patient understand? Will they know how this influences their prognosis? Will they agree with my decision on what is best?* These practitioners find the degree of uncertainty too great to bear, so they seek to minimise it by making decisions *for* their patients, as opposed to *with* them. At times this may be helpful to a patient; at other times, it may lead them to feel silenced by their doctor and frustrated by the lack of transparency – this depends, in part, on the patient's own tolerance of uncertainty.

Healthcare providers who are less tolerant of uncertainty also tend to order more diagnostic tests (such as medical imaging, blood tests and so on) or inappropriately refer patients on in an effort to squelch the anxiety of diagnostic uncertainty. This anxiety can come from a desire to reassure patients about their condition or from the healthcare provider's own uncertainty intolerance. Maybe a physiotherapist is seeing a patient who is complaining of lower back pain. He had a patient present with similar symptoms a few years back and missed that the pain was worse at night, a symptom that can be related to a tumour of the spinal cord – which, it turned out, was what the patient had. This prior negative outcome may result in a decreased tolerance of uncertainty in this current patient scenario, causing him to order an MRI (magnetic resonance imaging), get a second opinion and so on – just 'to be sure'. These steps, however, can be both expensive and delay critical progression in patient care. A 2022 article by Ryan Levi and Dan Gorenstein for NPR (US National Public Radio) examines this phenomenon in healthcare practice – particularly in the United States, given the healthcare system has few boundaries on how many and which tests can be ordered, if the patient can afford them. The article notes that 'MRIs done early for uncomplicated low back pain and routine vitamin D tests "just to be thorough" are considered "low-value care" and can lead to further testing that can cost patients.' Think of a patient waiting for surgery in end-of-life care: these tests can cause significant delays and may not even be relevant.

The irony is that these attempts to eliminate uncertainty can introduce new ambiguities without reducing existing ones. An MRI, which the physiotherapist considers critical to figure out 'what is wrong' with the patient, can generate even more questions. Levi and Gorenstein highlight a case with a patient who came in for an abdominal hernia repair:

Dr Meredith Niess saw her patient was scared. He'd come to the Veterans Affairs clinic in Denver with a painful hernia near his stomach. Niess knew he needed surgery right away. But another doctor had already ordered a chest x-ray. The test results revealed a mass in the man's lung. 'This guy is sweating in his seat, [and] he's not thinking about his hernia,' Niess said. 'He's thinking he's got cancer.'

… Though ordering a chest x-ray in a case like this was considered routine, Niess understood something her patient didn't. Decades of evidence showed the chest x-ray was unnecessary and the 'mass' was probably a shadow or a cluster of blood vessels. These non-finding findings are so common that doctors have dubbed them 'incidentalomas'. Niess also knew the initial x-ray would trigger more tests and further delay the man's surgery.

In fact, a follow-up CT scan showed a clean lung but picked up another suspicious 'something' in the patient's adrenal gland. 'My heart just sank,' Niess said. 'This doesn't feel like medicine.'

This patient has undergone another test, which involves exposure to radiation, and experienced stress due to the uncertainty around the source of this mass, to fulfil routine preoperative procedures – and the result was more questions than answers. Levi and Gorenstein note that the patient was finally cleared for surgery eight weeks after he first presented. Dr Niess wrote about the case in an article for *JAMA Internal Medicine*, dubbing it an example of a 'cascade of care': 'a series of seemingly endless medical tests or procedures'. There are multiple examples of this across the healthcare industry, such as prostate-specific antigen (PSA) testing leading to false positives or unnecessary treatment, and degenerative changes of the knee leading to low-value surgical interventions such as knee arthroscopy for osteoarthritis.

Of course, being tolerant or intolerant of uncertainty isn't a constant state of being and is case dependent when it comes to ordering tests or doing medical procedures. If, for instance, a healthcare professional who is normally tolerant of uncertainty spots multiple symptoms of concern, such as unexplained weight loss, fevers and night pain in the patient with lower back pain – which, combined, can be suggestive of an insidious cause such as cancer – tests are an appropriate response to detect a potentially life-threatening illness. Healthcare workers' tolerance of uncertainty, and responses to it, may change with each patient and in each context. The issue is if a healthcare worker is less open to accepting the presence of uncertainty or views it only as a negative.

The uncertain patient

As modern healthcare moves towards embracing the patient's autonomy and agency, and encourages them to be participants in their own healthcare decisions, the patient is an increasingly important source of uncertainty.

A participant in my team's research stated: 'Medicine is perceived as more black and white from the outside than it actually is on the inside.' This is a telling observation, as many patients come to a doctor and expect a diagnosis.

When healthcare providers should share uncertainty with their patients, and how much, are questions that some consider in need of more research. But when we reflect on my team's findings about the development of uncertainty tolerance in learners, alongside Han, Hillen and Strout's body of work around disclosing uncertainty to patients, we find several commonalities. This is not so strange. At their core, healthcare and education both rely on communication between a bearer of knowledge and a receiver.

A recent systematic review published in 2021 in the journal *Health Expectations* suggests that being transparent about uncertainty and the related discomfort it brings can help prepare patients for the uncertainty intrinsic in their care. Once the patient is primed for the impending uncertainty, the healthcare provider can describe a plan to address this uncertainty, including what steps the patient and healthcare team can take in the short term and long term to try to 'close the loop' on the unknowns – where possible. The patient's illness may not yet have a name, but the first steps in managing it are clear. This can provide the patient with hope and a sense of purpose in navigating the uncertainty, with their healthcare provider as their guide.

In Sally's case, this could involve the doctor reassuring her: 'Sally, we aren't sure what is going on. I can imagine that this is a scary time, but we have a plan to try to figure this out. I have consulted with my colleagues, and what we think is happening is something related to your heart because some of the blood results are on the high end and could indicate heart tissue damage. Your chest pain is also suggestive of this. We are here to help you through this. We are going to take another set of bloods to explore whether there has been a change. If they are higher for the heart-tissue-damage markers, this will suggest a stronger link to cardiovascular causes. The physiotherapist has also reviewed your chart and doesn't think the primary cause is a muscle strain, but they will be coming to see you to do some further examination to see your movement and pain in person. The parts that don't fit the puzzle currently are that you are relatively young and very active – but this just means we need to explore some more, and with your feedback and continued support, we will work together to figure this out. What questions do you have? How can I help you navigate this?'

The doctor is preparing Sally for the uncertainties ahead, explaining why the uncertainty is present and sharing her plan of action to address them. Instead of responding with, 'I don't know what is going on' or dismissing Sally's symptoms as anxiety or, worse still, saying nothing at all, she is being transparent, inclusive and aware of the uncertainty. She is leaving room for Sally to express her fears by inviting open-ended questions. The reassurance isn't that she 'knows the answer', but that they are on the journey together.

Ultimately, the extent to which a healthcare provider communicates uncertainty will depend on the specific context of each patient – and thus time spent exploring the patient's comfort level with and desire to learn about the uncertainties *before* sharing them without inhibition is essential. Just like educators who teach based on learners' prior knowledge and experience, healthcare providers need to explore the patient's capacity for managing uncertainty and adjust their communication style accordingly.

Associate Professor Francine Marques, who is based in the Faculty of Science at Monash University, knows firsthand how influential uncertainty is as a patient, and she understands why healthcare providers who learn to manage this inevitability can enhance the patient experience and patient care. Seven years ago, Marques was diagnosed with and received treatment for ovarian cancer. Some would think that, in having made it through treatment, she was through the worst of the uncertainty about her health. Yet it was not so: each follow-up appointment is itself an overwhelming source of uncertainty. The anxiety builds simply when she sees the date approaching in her calendar. The impacts of this uncertainty are palpable: 'One would think that after so long I would be used to it. The reality is quite different – you never really get used to even the thought of having cancer again. While consciously I may appear calm, I still notice headaches, back pains, insomnia and other symptoms of the anxiety creeping in the week leading up to my tests or appointments.'

After each appointment comes the waiting for the results. The uncertainty around whether the news will be good or bad is almost too much to bear. She begins viewing common physical changes as signs that the cancer has returned. 'Ovarian cancer symptoms are common in most people – women and men. It is easy to justify these symptoms as being cancer again, instead of most likely causes, such as something you ate.' Francine plans for surgeries and chemo and purchases PJs for the upcoming hospital stay while waiting for her testing results. She experiences symptoms of stress such as a lack of hunger, abdominal pain, nausea and sleeplessness. Suggesting she just 'become tolerant' of this uncertainty is unreasonable, and arguably dangerous for her wellbeing. Indeed, this form of uncertainty is so common that it has been given a name: 'scanxiety'.

Supportive care and acknowledgement of the emotional and physical toll of uncertainty on patients can enhance the patient's care experience and increase their capacity for managing uncertainty – just as it does for learners in the classroom.

Some healthcare systems, though, prevent this individualised approach to fostering uncertainty tolerance in patients. Instead, patients are treated like they are on an assembly line where every patient is the same and a single approach works for everyone. The result? Patients are less able to deal with the uncertainty, less likely to adhere to the healthcare recommendations and treatments, and more likely to feel hopeless, overwhelmed and lost. The assembly-line approach to healthcare pretends that uncertainty isn't present, as most healthcare systems are structured to suppress or ignore uncertainty.

How the pursuit of certainty impedes healthcare

In both Sally's theoretical case and Marques's lived experience, the risk to life factors heavily into how they feel about the uncertainty around their health status. My research found that most healthcare systems also tend to treat uncertainty as negative. Because healthcare often involves situations where a human life is at stake, a healthcare system (and a healthcare provider) can of course never be entirely tolerant of uncertainty, but there is a balance to be struck between the acceptable knowns and the risk with unknowns. One of my research participants describes the tension between uncertainty and risk to patient life in healthcare:

> [Healthcare providers] need to be confident enough to know that [they] can identify the people who are sickest and get them the right help in the right order. And, also, to feel a bit comfortable in the not really knowing – but not so comfortable that [they]'re blasé about it. [They need to think]: I'm probably going to need to go back and check that man in twenty minutes if he hasn't been taken inside, because I'm not sure what's wrong, and it could be something significant that will have a negative impact on his health if we don't do something.

Healthcare providers spend their days navigating the space between risk and uncertainty, whether they are consciously aware of this or not. Most healthcare systems tip the balance towards risk aversity: driving for standardisation and pushing for homogeneity and consistency in care to foster a sense of 'certainty'. This creates a structure that is intolerant of uncertainty. When we consider the significant risks they are designed to manage, uncertainty intolerant healthcare systems may be appropriate – until they aren't.

My team's research highlights the approaches that healthcare systems use to suppress uncertainty. These include checklists, standardised operating procedures and diagnostic normalisations. A nurse describes how these work in practice: 'The checklist, the tick-box, that most nurses are quite obsessed with, is all about feeling in control, which is all about counterbalancing the anxiety that results from the uncertainty and the unpredictability of a person['s] health.' The reality is that these measures are sometimes illusory attempts to control uncertainty. There are many reports of providers 'ticking the boxes' but failing at adequate patient care. The example of ordering the chest x-ray for a patient in hospital for an abdominal hernia repair represents this type of response to uncertainty.

A checklist is a noble attempt to generate certainty but it can also introduce ambiguities. This is exemplified by a radiographer's description of how x-rays are evaluated for diagnostic quality (quote is paraphrased for readability):

> If an x-ray is taken, the checklist is meant to evaluate whether that x-ray is of diagnostic quality. There might be 1000 different interpretations of that question, if you asked 1000 different people.

I work in breast screen. That's an area that's quite fine-tuned, as in the x-rays need to be good. The number-one reason for misdiagnosis is positioning of the radiograph. So, it's not that the radiologist's doing the wrong thing, or the wrong name is being put on the x-ray, or any of the 100 different things that could be wrong – the number-one reason for misdiagnosis is patient positioning. So, this has a significant impact on patient outcomes.

In breast screen, you have all these criteria that you have to follow. You grade each x-ray that you take. There are four mammogram images for each patient and there's all these different criteria. Is the nipple in profile? How much pec muscle is visible? What level is it? Have you got equal distance, equal amount of breast tissue on each image? … all these different things. You have to look at them all, and you have a very short period of time to look at these images and rate them.

When reviewing these images, there's two questions. Is this diagnostic? Does this meet the criteria that denotes good positioning? I understand why these questions are important, so that you can try to get the greatest majority of images at the best quality. As a radiographer, your images get reviewed by someone else every three months. And if you don't meet 50 per cent 'good', you get called up and you get in trouble, basically. So, it's a constant pressure that you need to get a certain amount of images meeting the standards.

But there's so much ambiguity. Sure, they give you the criteria, but like, seriously, did the peer reviewer see this patient? Or when the x-rays are marked 'below standard' – does that really impact diagnosis? Like, the nipple is 'not quite bisecting the skin edge', but it's so close I think the radiologist isn't going to think that's a cancer lesion. Despite this having no impact on the patient diagnosis, this infraction downgrades your evaluation score.

At the end of the day, it should be about how it impacts diagnosis or not. But you'll have radiographers sit there and argue about whether this is a good or whether this is a moderate image rating. So, there is still ambiguity within the rating system. And part of that is because everyone's breasts are different. In other cases, the images are fine according to the positioning criteria, but when you look at the images they are terrible and fail to help with diagnosis.

This comment underscores the unintended consequence of such checklists: they often fail to remove the uncertainties and, at their worst, can compromise effective healthcare. Here, radiographs may be technically excellent when evaluated through standardised criteria, but despite their high score, in some cases they could end up actually hindering the healthcare provider's ability to identify breast cancer – which is the whole point of the radiograph to begin with. Ultimately, humans are still the ones making decisions and interpreting the findings of medical imaging, which in and of itself introduces uncertainty that the healthcare system largely fails to acknowledge. Healthcare is, after all, a human endeavour.

AI and automation bias

Those in the healthcare sector who are aware of this disjuncture are looking for ways to manage uncertainty adaptively. The drive to minimise uncertainty in healthcare systems is so pervasive that the sector has robustly adopted and incorporated artificial intelligence (AI) into the workplace. AI excels at reproducibility and standardisation – which sounds like music to the ears of those in charge of a healthcare system fixated on consistency and minimising unknowns. When we dig deeper, though, AI's dependency on prescribed variables and emphasis on pattern recognition to generate recommendations means that AI is very much *unable* to deal effectively with uncertainty – the uncertainty which we know is at the centre of healthcare practice.

AI relies on available datasets, or sets of 'norms', to generate outcomes. This data is derived largely from 'capturable' healthcare statistics – datasets, for instance, that are input into electronic medical records (EMRs). The AI doesn't have access to all of the patient information in the healthcare provider's mind. Many EMRs sort patients into neat categories in relation to race, gender and so on, but these don't represent the actual specifics of the patient's case. When the AI is built from data acquired through healthcare research, this data often excludes diversity. Critics have pointed out that healthcare research has traditionally relied heavily on an evidence base of Caucasian, cis-gendered adult males. Such subjects then become the 'standard patient' against which everyone is measured. What happens when the patient doesn't fit this norm?

Medical and scientific journals are rife with reports that patients who aren't programmed into the AI as the 'standard patient' are receiving poorer healthcare and experiencing reduced healthcare outcomes. This inability to tolerate uncertainty, to manage that which the AI wasn't programmed to do, is deepening health inequities. For instance, a recent study published in the world-leading journal *Science* found that an AI algorithm was more likely to offer much-needed care to white patients than to Black patients despite the patients' needs being equal. This same

AI system is currently deployed to 'support' patient care for over 200 million people in the United States each year.

A September 2021 article in *Health Affairs* notes a similar bias, observing that 'even though Black Americans are four times more likely to have kidney failure, an algorithm to determine transplant list placement puts Black patients lower on the list than White patients, even when all other factors remain identical'. Katherine J. Igoe, writing on the Harvard School of Public Health website, notes that 'the Framingham Heart Study cardiovascular risk score performed very well for Caucasian but not African American patients, which means that care could be unequally distributed and inaccurate'.

In homogenising the patient population and erasing difference, healthcare systems create an 'average consumer' in opposition to the diverse and variable real-world healthcare population, and in so doing can create risks to patient safety (as opposed to minimising them). In a twist of irony, by trying to make all patients equal, healthcare systems are creating inequality. And AI, based on this flawed model, is entrenching the bias.

The *Science* article goes on to point out that most AI programming is proprietary – meaning doctors aren't able to question or evaluate whether the AI is fit for purpose. The uncertainty in AI comes from how the program is developed and what datasets it draws on. Given that the programming is what leads to these health inequities, *this* should be a source of uncertainty that healthcare systems work to suppress.

Scarily, AI is embedded and trusted across nearly all facets of healthcare despite being markedly intolerant of uncertainty. It is now engaged in nearly all healthcare contexts, including diagnosis and detection of illness, triaging patients, feedback during surgeries and clinical note-taking. Doctors are trusting the AI to make decisions about their patients and healthcare insurers are relying on AI to make decisions about whether a treatment is warranted or not. When you consider the depth and breadth with which AI is integrated into modern healthcare, the negative impacts stemming from erroneous standardisations are potentially enormous.

Compounding the health inequities perpetuated by AI is something known as 'automation bias', where people tend to trust machines as 'correct and certain'. Healthcare providers are more likely to fail to spot important patient irregularities missed by AI or to ignore contradictory information that challenges AI outputs. We saw this with the doctor who chose to trust the rapid flu test over their own intuition. The only certainty with AI is that the computer isn't right 100 per cent of the time, but there is a strong human inclination to trust it anyway.

Automation bias led to a Kafkaesque bureaucratic tangle for Sydney-based writer and early education and care advocate Lisa Bryant and her daughter, Rikki. Rikki became permanently disabled seven years ago and, following a lengthy application process, was granted disability pension, along with income protection. Because of a glitch in the computer system, the back pay Rikki received when her application was approved resulted in her being categorised as 'earning too much', causing her then to be cut off from the disability pension for making too much money. The people at the call centre didn't challenge the computer-made decision. In fact, Lisa writes, the 'human call centre is insisting it's right' – despite the overwhelming logical evidence to contradict this.

Another example includes when doctors rely on AI recommendations to prescribe medications for patients. Much of the research that highlights the value of AI in electronic prescribing suggests it should play a 'supportive role' to help doctors know when a treatment could be dangerous or ill advised. The AI recommendations are based on population-wide data – not on the individual patient being seen by the doctor. Consider a patient on antidepressants who has just found out they are pregnant. There is evidence that a patient with this profile is at higher risk of postpartum depression, so the doctor decides to recommend that the patient continue on antidepressants. The doctor types this into the EMR and an alert pops up that says this medicine is not approved for use in pregnant patients. The doctor researches this and finds out it is because the medicine isn't well studied in pregnant women. This alert is based on flawed population data. However, now the doctor must put a note in the chart that says they are not complying with the AI alert. Would all doctors feel comfortable doing this? The doctor is meant to make a final call, with the AI as a support tool, but in reality, the AI recommendations may come to influence the doctor's decision-making. Add to this the complexities around insurance and Medicare coverage, and the role of AI versus human gets even muddier.

These two cases illustrate how healthcare systems are set up to be intolerant of uncertainty. There are often system-wide protocols in place that prevent workers from easily overriding the electronic preferences, even when they are clearly wrong. The dangers with AI include both the trust that is placed in it and the speed with which it works. Thus, the human capacity for uncertainty tolerance may be just what the doctor ordered for working alongside AI in the healthcare system of the future. After all, it isn't the AI that is held responsible for mistakes: it is the human.

Diagnosis: burnout

Over and over, my research uncovered reports of healthcare providers who felt burnt out due to working in a healthcare system that continuously attempts to suppress uncertainty by using standardisations. They are frustrated by adhering to procedures that fail to support their diverse patient populations, by an overreliance on rigid guidelines and by the belief that healthcare decisions can be standardised and reproduced in any context. They are frustrated by the lack of acknowledgement that humans (with all their uncertainties) are still the ones, by and large, responsible for making the decisions, interpreting the findings and interacting with the patients.

This is not a new phenomenon. Seven years ago, the peer-reviewed medical journal *The Lancet* called out the disconnect between healthcare system administrators and those on the front line working with patients: 'Physicians, disillusioned by the productivity orientation of administrators and absence of affirmation for the values and relationships that sustain their sense of purpose, need enlightened leaders who recognise that medicine is a human endeavour and not an assembly line.'

Burnout is a real and present danger to healthcare systems. Dr Hester Wilson, a GP and a specialist in addiction medicine, regularly sees patients with complex addiction and mental health problems. In an interview with journalist and fellow GP Evelyn Lewin, she describes a patient with mental-health issues who had been bounced around the hospital system without receiving the support he needed. 'In the end I sorted something out,' she said. 'But it was at that point that I went, "I just can't do this job anymore. I just cannot do this. This is too hard."' Citing a 2019 Canadian study that suggests GPs struggle with 'frustration, exhaustion and compromised job satisfaction', she emphasises the relationship between burnout and uncertainty intolerant health-care systems:

> What they found is that GPs feel overwhelmed by the complexity of trying to manage their patients and their inability to actually assist this group of patients. And burnout is one of the things that happens as a result of that. It makes absolute sense that it gets to the point where as a human being you go, 'I just can't give anymore here. I can't manage this anymore. The negative emotions are too distressing. I've got to opt out.'

Dr Lewin points out that this correlates with a 2021 study in *The New Zealand Medical Journal*, which found that 45 per cent of senior medical professionals interviewed across a range of specialities were experiencing high levels of burnout. 'Burnout – a syndrome that is characterised by emotional exhaustion, de-personalisation and low sense of personal accomplishment – has been associated with a higher frequency of medical errors, lapses in professionalism, impeded learning, problematic alcohol use and suicidal ideation,' the study's authors wrote. 'Burnout is important because it can damage doctors and impair patient care.' Burnout has also been linked to a decrease in empathy, which can have particularly negative implications for patient care and outcomes.

My team's research suggests that one crucial factor driving this burnout is the gap between the intrinsic uncertainty in effective patient care and the system-level processes and procedures aimed at suppressing this uncertainty. In addition, over time, standard-isations and checklists negatively moderate healthcare providers' capacity for managing uncertainty, generating an intolerance of uncertainty across the healthcare system.

The healthcare system isn't the only source of uncertainty intolerance for these practitioners. Healthcare providers struggle with the emotional turmoil that comes with not being able to conclusively diagnose and care for a patient. There is also evidence that burnout rates and stress are particularly high among junior doctors. This suggests that burnout may, in part, be linked to doctor training: a significant number of new doctors are emerging from medical school ill prepared to manage the diverse and plentiful sources of uncertainty in healthcare settings.

A spoonful of sugar helps the uncertainty go down

So we have a diagnosis. How do we go about developing a treatment plan for managing the uncertainty in healthcare and building tolerance? A holistic plan requires changes across all tiers of healthcare: the healthcare system, the provider, the patient and the educator. We have to think big in order to see changes that will make a difference to patient care, so let's start with the healthcare system.

How to treat the healthcare system

The administrators of healthcare systems must refocus their approach from rewarding regulation and control to acknowledging the irrepressible uncertainty of caring for humans and building that recognition into systems and processes. Healthcare systems with built-in adaptability can be more accommodating, supporting patient care instead of hindering it, and can potentially reduce patient-related burnout among the healthcare workforce.

Person-centred care is built, in part, around the idea that one-size-fits-all healthcare can be a useful starting point but is not the end point. Person-centred care means focusing on the patient's humanity, which by definition is filled with uncertainty. Moving to flexible standardisation may be the key to balancing the uncertainty intrinsic in healthcare and the risk management administrators are concerned about. For instance, checklists that outline key questions to ask a patient during their first visit could use more accessible language (lay terms rather than anatomical terms, catering for those with lower levels of health literacy) and build in cultural context where this information may be of use in assessing a condition and how best to treat it. A healthcare provider could have the freedom to decide which of these questions to ask and the order to ask them in. Not every question on a patient history inventory is relevant in every situation, and too many questions can sometimes be unhelpful for a patient, overwhelming or confusing them, or raising new fears.

Healthcare systems also need to be designed to support the healthcare provider as the lead in patient care. Human healthcare teams could be built into workflow. For instance, following a series of medical investigations and AI recommendations in a complex case, human healthcare teams could meet to discuss the data and determine which data will contribute to the next steps of patient care, and to what extent. In this way, the AI is actually supporting the healthcare provider – but human judgement is prioritised. AI literacy training, professional development that educates about automation bias, and opportunities for healthcare workers and patients to co-create healthcare AI alongside developers, will go a long way towards balancing uncertainty with effective patient care.

Administrators have the opportunity to build a tolerance to uncertainty into the system by seeking creative human solutions to healthcare challenges and balancing human opinion with numerical rubrics. This could be done by engaging humans from diverse areas (including patients and other groups who engage in healthcare) and specialties (think tanks), to discuss systems and processes within healthcare. Instead of relying on patient-satisfaction scorecards, which focus only on the end point or outcome, working groups would help develop the processes affecting their healthcare journey, from the way patient waiting rooms are set up to the time and types of interactions, and so on. Instead of the healthcare system being exclusively designed by administrators and providers (and AI), community partners – those interfacing with the system – would be part of this process.

Managing the uncertainties of healthcare against the risk to human life is challenging and at times can overwhelm even the most experienced healthcare practitioner. My team's research found that the simple act of sharing and discussing uncertainty with peers helped raise awareness of its universality. Providing a formal way to acknowledge uncertainty in healthcare can help individuals to engage with it productively. Healthcare systems should be designed to incorporate more support networks to discuss sources of uncertainty in patient care and have scope for more debriefing sessions that focus on sharing the stress and anxiety – the underacknowledged emotional tax – that managing uncertainty can bring. Such services could be targeted both to providers and to patients. This might not sound cost-effective, especially in the context of what in many countries is an already overstretched health system. But with media reports claiming that, for instance, nearly one in five US healthcare workers have quit their jobs over the last two years due to burnout and uncertainty about staffing, salaries and the Covid-19 virus itself, establishing ways to deal with and talk about uncertainty could pay dividends in terms of staff retention and welfare, and ultimately in terms of improved patient outcomes. It should be a priority for administrators of the healthcare system.

Mistakes will always happen; the unexpected will occur. Blaming individual caregivers for mistakes after the fact is unhelpful and can undermine their capacity to be tolerant of uncertainty. Furthermore, designating an individual as the source of a mistake is dangerous and misleading because it implies that this is unexpected in a complex system. The blame game simply results in people *not* speaking up due to fear of retribution, even when there is an obvious mistake occurring that puts patient safety at risk.

Studies have shown that a healthcare culture focused on safety, instead of blame, works best to support patients. If healthcare teams feel that they can speak up, effective solutions are more likely to be found. In 2006, emeritus professor of surgery at Drexel University (and my father-in-law) Dr John Clarke wrote a review focused on changes to the healthcare system to improve patient safety. A key finding was that the culture of the system has a significant impact on reporting, and addressing, medical errors.

Dr Clarke, a leader in the state of Pennsylvania's patient healthcare safety review, found that 'a lot of medical errors result from uncertainty in the information being communicated':

This is usually the result of a combination of 1) different people having access to different sets of information and 2) information changing over time, sometimes very quickly. This combination of disjointed information can result in different healthcare providers having very different ideas of what is happening with the patient at any given time. Different information is known to different people at different times resulting in, for instance, a surgeon might have one mental model, while the anaesthesiologist has another. Effective communications supporting reliable healthcare delivery requires that everyone shares the same goal, has the same mental model, and has the same level of situational awareness – background, the assessment, and the recommendations for action.

To address these communication challenges, he suggests that healthcare workers write down information as it comes in; suggest to patients to bring along an 'advocate', especially when stressful news is expected; support patients to develop their own personal health record by making medical records accessible and available to them; ask active, open-ended questions instead of leading ones (for example, 'What is your name?' versus 'Are you Mr Jones?'); calling out information to the team ('thinking aloud') so that the whole team has the same information, and have the team repeat it back; and, of course, double-checking all important information (for example, where the pain is, or the location of the surgery).

An example of uncertainty in communications is the practice of colour-coded wristbands to indicate high-risk patients or those with conditions such as allergies. In 2005, a hospital in Pennsylvania

submitted a report to the Pennsylvania Patient Safety Reporting System (PA-PSRS) describing an event in which clinicians nearly failed to rescue a patient who had a cardiopulmonary arrest because the patient had been incorrectly designated as 'DNR' (do not resuscitate). The source of the confusion was that a nurse had incorrectly placed a yellow wristband on the patient. In this hospital, the color yellow signified that the patient should not be resuscitated. In a nearby hospital, in which this nurse also worked, yellow signified 'restricted extremity,' meaning that this arm is not to be used for drawing blood or obtaining IV access. Fortunately, in this case, another clinician identified the mistake, and the patient was resuscitated.

This single near-miss report prompted the Pennsylvania Patient Safety Authority to explore this further. The organisation surveyed all hospitals in the state to identify which wristband colours were used to indicate critical patient information. It turned out, to quote Dr Clarke, that Pennsylvania hospitals used a 'kaleidoscope of colours'. A pilot was started by a small group of hospitals in northeastern Pennsylvania using standardised wristband colours – to great success. The result? The entire state now has a standardised colour system to convey patient information. This has been expanded across many hospitals in the United States – with purple denoting 'do not resuscitate'.

Uncertainty is ever-present in healthcare, and by developing a culture of safety (as opposed to fear), unnecessary ambiguities can be reduced so we can focus on the relevant uncertainties – those that brought the person into the hospital to begin with.

Rethinking the 'standard patient'

Many evidence-based practices (where research data informs and guides patient care) are based on a 'standard patient' who is white, straight, cis-gendered and male. This does not reflect the diversity of patients in the community. Governments use this research based on the 'standard patient' to develop policies and standardised operating procedures – but if the 'standard patient' is in fact not the typical patient, and the majority of patients (with their unique variables) are excluded, what value does this data add? How reliable are policies developed for only a subset of the population? The administrators of healthcare systems, and we as a society, need to ask of the findings, 'Normalised to who? Normalised to what?'

Take the number of Black ciswomen who are banned from competing in the Olympics due to their normal levels of testosterone, which happen to be higher than the standards set as 'acceptable' for women. Or the evaluation of health through the BMI (Body Mass Index). What many may not know is that the rating of 'healthy' or 'unhealthy', 'normal' or 'overweight', stems from a predecessor measurement designed exclusively for Western European men.

The contemporary BMI chart, now used widely in healthcare and health apps, isn't much better. The weight categories and related health status include metrics of men from different geographical location – but still only metrics of men. This calculation fails to accurately assess 'health' for many, including elite athletes (who can be inappropriately labelled as 'obese') and those of Asian descent (who tend to be inappropriately classified as 'underweight'). Women also lose out in the BMI calculation because the older we get, the better it is for us to carry a bit more weight – but if we do, the more likely we are to be classified as 'unhealthy' by BMI standards. Many women who have multiple markers of health are labelled 'unhealthy' solely based on this (to quote a 2022 *ABC News* article) 'really old, racist and non-medical' measurement. Yet the notion of 'normal' BMI persists across healthcare settings. By asking who and what the norms are defined by, we can begin to provide boundaries about certainty and uncertainty in a way that can improve our understanding of health.

The sparks of change are beginning to ignite at last, and we should all encourage them into flame. A February 2022 article in *Health Affairs*, a journal focused on health policy, suggests that race and ethnicity should consistently be included in data reporting. It highlights three areas where the lack of this information may be contributing to further disadvantaging certain patient populations: patient feedback and perceptions of healthcare, patient safety reporting data and patient safety outcomes communication. For instance, they note that 'studies have demonstrated that Black patients consistently experience higher rates of hospital-acquired illnesses and injuries during surgical procedures than white patients'. The scientific journal *Nature* now requires mandatory reporting of sex and gender in all articles published under its banner. It is a small but important first step. Healthcare systems should take into account relevant information about populations and their key characteristics in order to improve patient care. Instead of focusing on inflexible standards, transparent reporting about where uncertainty exists in this data (that is, which populations have not been studied) can support healthcare practice more broadly.

STRATEGIES TO MANAGE UNCERTAINTY AS A HEALTHCARE ADMINISTRATOR

- Create flexible standards that balance the need to reduce risk with recognising uncertainty. Such standards could allow for a better assessment of patients and their individual treatment needs.

- Develop a model of patient care that cultivates interactions between specialities and across health professions. This fosters adaptation and creativity in response to complex health problems and ensures the best patient outcomes.

- Move from a culture of individual blame to a culture of collective patient safety that acknowledges uncertainty as part of complex healthcare practice.

- Rebalance the relationship between human and AI decision-making in healthcare and create opportunities for healthcare providers to play a role in developing the technology for use in their fields.

- Support research and data collection in patients with complex healthcare needs (such as multiple medical conditions) and patients that are underrepresented. Identify when the data isn't relevant to a given a patient due to these limitations, and act accordingly.

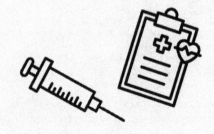

How healthcare providers can adapt to manage uncertainty
Healthcare providers sometimes find themselves stuck in the middle between the patients' expectations and the systems within which they work. There are steps, though, that providers can take to help manage healthcare uncertainty even with external pressure to the contrary.

An important first step is reminding yourself about your purpose and role in navigating uncertainty in healthcare. Your primary motivation may be to help ease the patient's burden, to be excellent at your work, or simply to make it through the day – whatever it is, reminding yourself of this purpose when the uncertainty builds is useful. Next, sharing uncertainty when you see it – with trusted peers, with your patients (as appropriate), with your support network and with your staff – not only helps you manage the unknowns; it also helps others to embrace uncertainty. You can help others realise that uncertainty isn't the exception, but rather the rule.

Once you are aware of unhelpful sources of uncertainty in a patient's case, you can take steps to minimise them. When you have gathered as much information as is reasonable, find people who can discuss the patient's care with you from a unique point of view: consult specialists in other fields, nursing staff, peers and so on. Forming a diverse team will pay dividends – while working with others who are different from you can sometimes introduce new professional uncertainties, it can also lead to innovative solutions. In addition, this collaborative approach means that the patient can feel reassured they are being served as well as possible by a team of healthcare professionals. The positive impact on patient care and outcomes will likely outweigh any short-lived uncertainty about your professional identity.

Lastly, embrace the presence of uncertainty. As a healthcare provider, you cannot control every outcome. There will always be curveballs – idiosyncrasies in the patient's physiology or their response to treatment; unexpected events in the treatment setting; irregularities in the medical imaging or diagnostic tools. If you can accept that inevitability, you will be better prepared to deal with the curveballs as they come – provided the system you work within is also tolerant of these uncertainties. Talk to your patients about the sources of uncertainty in their situation and clue them in on what to expect.

You can also use your experience as a practitioner to minimise uncertainty. You are skilled at being able to identify when the pattern doesn't apply, when the one-size-fits-all approach won't work. At times, you might also have an inkling that something is wrong with a patient, even if diagnostic tests or observations say otherwise. There is power in questioning what appears to be certain. Make the most of it when the situation calls for innovative thinking.

And PS: do pay attention to your own mental health. Those who are geared to help others sometimes overlook the need to mind their own health first. Remember that it is okay to take time out or sound the alarm about potential burnout. Particularly during a pandemic that has stretched healthcare resources and exposed many in the sector to unprecedented levels of professional and personal stress, gentle self-care can help to reconnect practitioners to their purpose and build a tolerance of uncertainty.

STRATEGIES TO MANAGE UNCERTAINTY AS A HEALTHCARE PROVIDER

- Engage in person-centred care: be transparent about sources of uncertainty in a patient's case and develop a plan to address these sources of uncertainty, explaining to the patient the purpose of each step.
- Avoid becoming overwhelmed by uncertainty by keeping clear on your sense of purpose and duty of care to the patient, and by sharing your experiences with colleagues.
- Help reform the healthcare system by being part of the change.
- Increase your tolerance of uncertainty in healthcare, and exploit this to challenge the scripts and standards.

How patients can adapt to manage uncertainty

As patients, we want our healthcare providers to be tolerant of uncertainty, to challenge the AI algorithms when they may not be serving us well and put the humanity back into healthcare. We also want a clear diagnosis, treatment plan and prognosis; our questions answered and fears assuaged; and a balance of emotional reassurance and straight talk. It's a tall order.

We too need to learn to embrace a level of uncertainty in relation to our health. Television depictions of healthcare, where everything is certain, quick and clear, rarely match reality. A definitive diagnosis isn't always possible. Even patients with the same illness may require entirely different healthcare management. And there is a whole lot more bureaucracy than *Grey's Anatomy* makes out.

Embracing uncertainty is a challenging task when we are unwell and feeling vulnerable – tolerating may be more accurate. How do we best go about it? We need to be open with our healthcare providers about our comfort level regarding uncertainty – how much information do we want, and how much is too much? When do we want (even need) greater help from our providers in making an informed decision?

We can ask questions and entreat our healthcare providers to discuss possible outcomes and solutions. If viable, it may help us to refrain from focusing on a single form of treatment or diagnosis as the key to success. We can also keep focused on outcomes. For instance, if a definitive diagnosis isn't possible, we can concentrate on how our symptoms will be managed. By decreasing the pressure on our caregivers to give us *the* answer in cases where this is not straightforward and focusing instead on the process of improvement, we allow for a more productive conversation and improved health outcomes.

Seek out a support network to be able to discuss and share the challenges of managing complex and uncertain health issues. Ask your caregiver to refer you to other forms of help and assistance. Bring a support person to appointments if it helps to have someone else who can absorb the information.

Together, these approaches can give us a purpose and plan as we balance the anxiety of our health with the uncertainties related to our care.

STRATEGIES TO MANAGE UNCERTAINTY AS A PATIENT

- Be aware that uncertainty is intrinsic to healthcare and that while you have every right to be informed about your health, in some cases a definitive diagnosis or prognosis may not be possible.
- Determine your own level of uncertainty tolerance and share with your providers how much and what sort of information about your condition you want to receive.
- Ask questions and maintain a sense of curiosity about treatment options, where it is reasonable to do so.
- Build a support team around you to help manage the negative impacts of uncertainty, including the anxiety of waiting for results.
- Acknowledge that the uncertainty around our health is challenging to many.

How educators can better teach uncertainty tolerance

The importance of uncertainty tolerance in healthcare education is evident across both the research and the sector. More and more medical schools and governing bodies list synonyms for uncertainty tolerance – such as 'managing uncertainty', 'tolerating uncertainty' and 'handling unknowns' – among the desired attributes of their graduates. Some are even calling for uncertainty tolerance to form part of the student selection process – that is, only those with a certain level of uncertainty tolerance (a level appropriate for managing the plentiful sources of uncertainty in healthcare) would be allowed to undertake healthcare courses.

My team's research, and the significant contribution Dr Georgie Stephens at Monash University has made to it, counters the decades-old concept that uncertainty tolerance is static and easily measurable. Our work suggests that the development of uncertainty tolerance can be influenced by educational approaches – so prohibiting prospective students from entering medical training before they have an opportunity to develop their tolerance of uncertainty may result in eliminating potentially excellent future healthcare professionals. Furthermore, reliable scales would be needed for such high-stakes decision-making – and a holy-grail uncertainty tolerance measurement tool does not currently exist.

In fact, testing potential healthcare students for uncertainty tolerance may simply highlight failures in the educational system. To get into medical school, students have to score highly on standardised exams where a single best answer is expected. In selecting for students who have done well on these exams of 'certainty', we may inadvertently be selecting for burnout in a future healthcare workforce. These students are geared to look for certainty (and they are rewarded for it), but their work will involve managing unknowns, for which they may be underprepared. This gap between expectations and reality may be a key factor in the psychological stress researchers Jason Hancock and Karen Mattick found in those who are intolerant of uncertainty.

Instead of selecting students who are more tolerant of uncertainty, we could use selection criteria that illustrate to prospective students that uncertainty tolerance is essential in healthcare. While standardised exams may have a place in selection processes, admissions processes need to reinforce that being able to balance knowns and unknowns and navigate uncertainty matters more.

In line with this thinking, Australian healthcare schools have been encouraged to diversify the benchmarking for student selection, as is common in the United States. Recent studies suggest there is value in considering students' grade-point average. This finding is repeated across multiple studies, contexts and cultures. Grades are typically based on more than performance on a standardised exam: they reflect a diverse range of assessment tasks such as reports, debates, group projects and so on that are collected over a given period of time. Grades may better demonstrate a healthcare student's capacity to adapt to unknowns and balance knowledge and ignorance – depending on the type of education leading to these grades.

The multiple mini interview (MMI), which is used in a variety of institutions across Australia, the United Kingdom and the United States, is another potential useful selection tool. A recent review found that MMIs are one of the strongest predictors of student performance in healthcare degrees. These short (five-to-ten-minute) interviews, in which a series of scenarios are presented, allow assessment of attributes such as propensity for teamwork, empathy and communication skills. They offer opportunities to assess potential students' abilities to manage uncertainty because each scenario is unknown and multiple approaches in responding to these scenarios can be considered valid.

In the classroom, there are a variety of strategies educators can use to foster uncertainty tolerance in learners and better prepare them for the uncertainties of healthcare practice. A curriculum that offers regular opportunities to practise managing uncertainties can further support future healthcare providers to develop uncertainty tolerance. The methods for introducing and fostering uncertainty tolerance in learners is outlined in

Chapter 1, but here are some pointers relevant to those studying healthcare.

It is important to introduce students to where and how uncertainty will be present in their future jobs, so that it's clear the skills they are learning in the classroom will have a purpose in a professional context – this could be done by telling the story about the flu to the class at the start, as I do with my students. Then, stimulate uncertainty and moderate it so learners begin to develop the types of thinking they will use for managing future clinical uncertainty. For instance:

- Encourage class debates around the ethics of healthcare. For example, is it appropriate for a doctor to refuse to perform a procedure, such as a hysterectomy, if they disagree with the patient's choice?

- Present clinical cases where more than one correct diagnosis is possible. This is sometimes referred to as 'case-based learning'. Typical clinical cases in healthcare education end with a definitive diagnosis. Cases that have more than one possible answer mimic the clinical environment more strongly. In the clinical environment, doctors may develop a list of differential diagnoses, which include one or more likely diagnoses, some diagnoses that could also explain the patient presentation but are less likely, and diagnoses that may be unlikely but are important not to overlook because they have grave consequences.

- Develop a progressive case study where students are given minimal information about a patient's symptoms (a high level of unknowns), and as the students work through the case, more information is provided. Some of this additional information is clear-cut, and some of it (such as diagnostic and investigative results) is open to interpretation.

My team and I use these approaches in our own medical teaching. Remember Sally and her chest pain? This is based on an example of a progressive case we often give to students. We start with a paramedic handover, where students learn Sally's chief complaint of chest pain alongside demographic information such as her age and gender. Students are tasked with working together in diverse teams (this type of teamwork is expanded upon in Chapter 1) to answer a series of questions, such as *What is your differential diagnosis, and why? What would you do next, and why?* After the teams hypothesise about what Sally could have, we let them decide what tests to order throughout the case – creating opportunities to manage the uncertainty of which tests to order and when. Maybe they choose to order the bloods and are left to interpret which results are relevant and whether the findings are significant. Is a high glucose level important? How high is 'high'? When was the last time the patient ate?

As students progress through the case and gather more information, they start to learn that uncertainty doesn't disappear as more information comes in – the sources of uncertainty are just different. Furthermore, because they are working in diverse teams, students are introduced to the uncertainties related to teamwork (what is my role in the team? How do I communicate with others?). These are all common forms of uncertainty healthcare workers encounter in the workplace. The educator checks in with each group's discussion and draws on a variety of moderators described in Chapter 1, including intellectual candour (explaining their own experiences with managing uncertainty) and giving the students purpose in managing this uncertainty (linking it to their future role in patient care).

As students progress through their healthcare education, the uncertainty tolerance pedagogy can change to meet the students' expanding capacity and the increasing demands expected of them. Simulations are an oft-used approach for helping prepare students for clinical placements. These traditionally focus on helping to develop students' technical skills, such as in CPR or how to place an IV, and so tend to focus on tick-box activities with clear outcomes and little uncertainty. My team's work suggests that with small tweaks, these simulations could also help students develop their uncertainty tolerance. If simulations are adapted to become destabilising experiential learning activities – that is, they have a variable or vague end point where a clear diagnosis is unattainable or multiple treatment routes are possible, and/or where the patient's communicated response is complex – these situations demonstrate the uncertainties typical of healthcare practice. Some examples of tweaks to simulations include making a patient's vitals change midway; making the patient's condition resist a clear diagnosis, encouraging students to focus only on management of their care (and not the cause); or making the patient highly stressed and anxious, challenging the students to practise communication strategies in the face of uncertainties. Another tweak may be engaging interprofessional collaboration, where nurses, doctors and allied health professionals work through a simulation together as a healthcare team. In the simulated destabilisation, students are forced away from thinking about the patient theoretically and are instead tasked with acting out this scenario in real time with others, just as in a workplace context. Many medical schools in Australia, the United States and the United Kingdom already use these types of uncertainty-infused simulations, but all should build it into their curriculums.

Clinical placements, typically the final stage of healthcare education degrees, extend students' opportunities to practise managing uncertainty – as clinical placements are filled with it.

This workplace-based learning is akin to the iso-immersion stage of the experimental learning spectrum discussed in Chapter 1. Crucial in this type of uncertainty tolerance pedagogy are the moderators educators use to help students manage the many and diverse sources of uncertainty. The level of uncertainty in healthcare settings can cause even a typically tolerant student to feel overwhelmed. Our data indicates that clinical-year medical students are overwhelmed by uncertainty: they are unsure of their purpose when on placements, unsure of what is relevant to learn, how to approach the learning, who to go to when they have questions, how to interact with patients and the healthcare team, and so on. On top of this, these students are also managing uncertainty related to their personal lives. Every few weeks these students are placed in a new ward, on a new rotation, and the uncertainty swells again. This uncertainty due to lack of familiarity with surroundings is also typical of the students' next phase towards becoming a registrar, as junior doctors will often undertake four to five rotations in a single year.

Given this, clinical educators would do best to double down on moderating clinical-year medical students' uncertainty tolerance through providing support and opportunities to critically reflect. This could include comprehensive first-day orientations where students are introduced to the ward, their role in the team, the relevant people to reach out to, and so on, to help place boundaries around unnecessary sources of uncertainty. A healthcare educator who engages in intellectual candour and shares how uncertain they felt when they first did their rotations can be reassuring to these students and help to reduce feelings of anxiety. Allowing dedicated time for these students to engage in reflective practice, through debriefing sessions or a chat over coffee, can also help them manage the plethora of uncertainty. Experienced supervising doctors can provide valuable pastoral care for students and can help to build a tolerance of uncertainty in students. Peer networks may also be powerful, as my team identified that peers and near-peers fostered reflective practices that helped increase healthcare students' tolerance to uncertainty.

STRATEGIES TO MANAGE UNCERTAINTY AS AN EDUCATOR

- Understand that uncertainty tolerance is not static but can change over time through targeted teaching strategies.
- Build in opportunities for students to 'practise uncertainty' in the classroom by tweaking activities and drawing on the strategies outlined in Chapter 1.
- Signal the value of uncertainty tolerance in the curriculum by designing assessments that value process over the end result.
- Integrate critical reflection, including through peer and near-peer interactions, to encourage students to discuss the benefits of learning to manage uncertainty.

The antidote to uncertainty

Healthcare uncertainty affects everyone at some point in their lives – whether as a provider, an educator or a patient. When we pretend that healthcare is an exact science where the correct diagnosis is always made, and clear, appropriate treatment plans always developed, we are fooling ourselves. Being transparent about uncertainty in healthcare (and the discomfort it can bring) and building a healthcare system that recognises uncertainty as intrinsic to medicine can improve patients' lives and healthcare providers' wellbeing.

Healthcare systems that treat risk and uncertainty as synonymous tend to focus on suppressing uncertainty through standardisations and rigid guidelines. While the intentions behind this approach may be well-meaning, uncertainty will always find a way to seep in. By suppressing uncertainty, healthcare systems may be both failing to reduce risk and introducing new risks to large patient populations who don't fit the mould. We need to flip the script and move from healthcare systems that are 'uncertainty avoidant' to those that are 'uncertainty aware'. This transformation will allow for appropriate flexibility to be built back into the system.

A more flexible, human-centred healthcare system creates an environment that fosters uncertainty tolerance in its workforce. The result? The potential for a workforce that is more stable, less burnt-out and more capable of equitable healthcare delivery, with an enhanced capacity to deliver collaborative, high-level care. This uncertainty-tolerant workforce is less likely to overspend on irrelevant tests or delay diagnosis, leading to quicker and more cost-effective outcomes. In this context, both patients and providers can build their capacity to deal with uncertainty – the latter with help from healthcare education, which builds in opportunities to practise managing uncertainty at an early stage.

What about Sally? She is stabilised and back at home. She still doesn't know the cause of her chest pain, which evaded clear diagnosis, and she may never know. But Sally can live with that uncertainty, reassured that her caregivers are monitoring her to keep watch for the risk of a serious heart incident. And that is the point.

3

COUNTING ON UNCERTAINTY

How those in economics and business
can deal with the unknown

'If there's one thing that's certain in business, it's uncertainty.'
—STEPHEN COVEY

Throughout 2022, we endured breathless media speculation about reserve banks and boards across the developed world and their decisions around interest rates. Tension was high: will they or won't they raise rates? How much inflation is relevant to act upon? When do cost-of-living pressures begin to bite? Experts were brought on to morning television and evening news broadcasts to discuss and debate potential rate rises. Many, including myself, tuned in, hoping for insights as we considered the ramifications of interest rate rises on our personal finances. But each broadcast panel seemed to have a different opinion – a consensus about the degree of economic uncertainty, and when it would lessen, was nowhere to be found. Viewers were left to decipher conflicting expert opinions when making choices about what to do with their household budgets. We are in unprecedented times, many experts said. Some admitted they couldn't offer predictions based on the past because the current situation is so, well, novel.

Their reluctance perhaps stemmed from the lessons of the recent past. The pandemic began with similar economic predictions from all the major players. Then, there seemed to be more certainty in the forecasts of our economic future. Experts worldwide appeared to agree that doomsday economic scenarios were the order of the day. Newspapers shouted headlines like 'Global economy could lose 1 to 3 trillion US dollars!' and 'Coronavirus to lead to biggest contraction of the economy since Great Depression'. What happened instead for the next two years was, for many developed countries, a period of better-than-expected economic performance, a housing boom and an increase in consumer savings. Those who still had jobs spent their savings on household goods and products, and businesses rapidly adjusted to online trading, accelerating digital commerce exponentially ahead of pre-pandemic trends. We still faced a heavy, devastating death toll from Covid-19 in many places, and some of the predicted economic challenges, such as supply-chain issues and workforce deficits due to lockdowns and border closures, did come to pass. But it was only as the pandemic restrictions appeared to be easing that the most protracted economic impacts began to be felt, surprising many.

Even positive financial outcomes generate uncertainty. The pandemic spurred a desire for better living conditions as an unprecedented number of homebuyers upgraded or entered the market for the first time. As housing prices rose, some turned to budgeting software and financial planners to help protect their personal financial security and manage their income in the event of interest rate rises.

How do we plan for the future when there are so many unknowns, both positive and negative, in how the economy will behave and in how reserve banks and boards will manage this? How can we plan for the unexpected: for the economic decisions that are out of our control, for the spontaneity of good fortune?

Economic and financial decisions at government level may seem removed from our daily lives, but in fact they affect the global wealth and the international standing of individual countries, as well as household survival and societal wellbeing within countries. The state of the economy has an impact on our employment, household savings, consumption level, investment choices and so on. Our personal financial wealth determines our ability to put a roof over our head and food on the table, but our national economy influences the cost of these items and our global security. So economic prediction is an important game. There is a strong desire for financial certainty and economic security in both the public and the private sector.

At the same time, most of us (understandably) have only a limited understanding of how economic management works and what skills it involves. A research participant described a common perspective on accounting: 'The community understanding of accounting, generally, is the field is often regarded as being focused on counting. It's often regarded as boring because it is just procedural. It is just about mechanics, and there is no opportunity for creativity or judgement. There's a perception that there is no need for critical thought.'

But numbers can be more complex than they first appear. Humans need to make decisions about the numbers: what equations to run, what questions to ask and what conclusions to draw about the future. Economics requires significant critical thought.

This is because it is anything but certain: it is driven by human behaviour, which is incredibly variable. The spending decisions we make, the values we hold, and our responses to risk and uncertainty often look different in practice than in prediction.

Number crunching – using the past to predict the future

The business and economics sectors attempt to satisfy our hunger for certainty, for managing contemporary unknowns and planning for future economic uncertainty, by 'predictive modelling'. By drawing on historical data, or contemporary data that closely resembles the predicted future context, these models attempt clairvoyance; they map an average or 'typical' response to a given economic uncertainty and make a prediction about the future.

Indeed, the early Covid-19-related economic projections drew upon modelling of datasets from what economists perceived to be similar previous scenarios, or from reasonable predictions of what was likely. Some chosen models drew on the economic experiences of the SARS and Avian flu pandemics; some were programmed to draw on data based on Covid-19 being confined to China (as had occurred with other viruses in the past) and predicted how an impact on the Chinese economy would influence the rest of the world, and so on. This modelling, which was undertaken amid a fast-developing situation where the information changed daily, made many fearful and concerned about the economic future in the face of a pandemic the world had never experienced and whose end point remained unknown.

Many of us also rely on predictive modelling in our approach to managing uncertainty in our personal finances. Programs designed to help us budget and save typically rely on our past spending habits. Open up any budgeting software and the first thing you are asked to do is input information about your recent purchases into the program. How much did you spend in the preceding three to six months? What did you spend your money on? What were essentials versus non-essentials? Based on your answers to these questions, the program creates a budget. The software is built based on a key idea: past data is highly likely to predict future spending. But what if the preceding time period wasn't predictive? What if the ramifications of the unknowns aren't considered? What if interest rates go up, or you lose your job? What if you get injured? What if your financial values change, and your spending changes with it? One probability few of us considered prior to the Covid-19 pandemic – what if you start working from home regularly? There are also the large-scale unpredictable unknowns that can affect personal finances: pandemics, climate change, wars and so on. These 'unknown unknowns' often generate economic uncertainty because they are novel, unpredictable and complex – all attributes of uncertainty stimuli.

While we hope that the historical data can predict the future, when the economic uncertainty is stimulated by novel unknowns, we may find this modelling to be more fortune-telling than predictive. The economic modelling for Covid-19 depended on prior pandemics. But how many pandemics had caused massive lockdowns where workers were barred from coming to work and began to work from home? How many caused long-term global health burdens similar to Covid-19? The Covid-19 pandemic has resulted in a very different toll on global economies than previous pandemics because the prior data couldn't guide us. The 'typical' pandemic economic pattern didn't apply. Ultimately, the very human choices we make (including everything from daily household spending decisions right up to government policies around spending, welfare and economic relief) introduce uncertainty that traditional predictive models are not programmed for.

Businesses are also affected by this logic. While profit may be king for many businesses, the social values that a business may have can influence the financial decisions it makes. Consider a textile company with owners who value sustainability – while the modelling would indicate that the greatest financial gain can be made through manufacturing their products offshore, the company's ethos means that a decision to use sustainable onshore manufacturing is made, to decrease the carbon footprint and support planetary health. If historical economic data is used, dating back to the Industrial Revolution, where sustainability was not typically considered, the predictions would likely not support the contemporary or future trajectory of this business.

Retailers use historical data to model consumer purchases. Our spending habits are mapped from the data that emerges each time we swipe our credit cards or loyalty cards. Vendors use this information to guide their decisions on which inventory to purchase and when. Typically, this modelling gives a reliable general overview of the average consumer spending patterns within age groups, income brackets or with the onset of different seasons. Yet many bookshops saw the limitations of this type of predictive modelling when the disrupter Amazon appeared. The modelling didn't have the power to predict the future impact that digital commerce would have. Bookstore owners may have assumed that the appeal of wandering down aisles of books, perusing, and chatting with friends over coffee would be preferable to online shopping. And it is true that many consumers still valued this experience. But thousands began to go online to purchase books. Many readers decided that the ease of online searches, the convenience of home delivery and/or the competitive pricing had greater allure than the physical environment of a bookstore. These models failed to predict the very human power of change and adaptation.

Though this isn't a story of woe. Bookstores around the globe also adapted. By facing the uncertainty, instead of ignoring or being paralysed by it, those in charge recognised the gaps. Some bookstores are experiencing a resurgence by curating a collection relevant to the local community – instead of aiming to be a 'one-stop shop'. They are catering to the natural diversity and complexity of their local neighbourhoods, something Amazon cannot do.

The same is true of global economies – predictions are often misaligned with what comes to pass. For instance, when the media was speculating on the uncertainty surrounding the Reserve Bank of Australia's interest rate decisions, some experts modelled their predictions on the recent interest rate decisions of Pacific neighbour New Zealand – as this was a reasonable known data source from a context that seemed similar enough to Australia's to help predict the future. Others focused on what the United States and the United Kingdom were doing. The uncertainty here is to what extent the values and direction of these global economies matched those of Australia. Indeed, several finance journalists pointed out that the Reserve Bank of Australia would be waiting for signs of local wage inflation (when national wages increase over time), whereas other economies may have acted with general inflation (when the purchasing cost of supplies increases, increasing the cost of goods). The accuracy of the predictions depended on more than numbers – it depended on whether these models were able to allow for the uncertainty of human decisions and values, and whether those running the modelling employed critical thinking to interrogate the results.

The challenge with predictive modelling comes when events don't match the norm, such as when the information is incomplete; when we disregard our very human emotions and behaviour, including panic and fear, or our deep-rooted needs for connection, which can influence our financial decisions; or when there is diversity in the population and few in the community are represented by the mean; and when the system is complex, as with a global economy.

The Reserve Bank of Australia needed to consider quite a few unknowns, which the modelling may not have accounted for, when making decisions in a pandemic context – in *this* pandemic context. Would an interest rate change alter lenders' practices around home loans? Would this have knock-on effects to the housing market, and the small businesses relying on robust house sales? Would an interest rate hike cause household spending to drop as people became fearful of further rises? Could this result in a negative impact on retailers, who then need to make decisions around staffing, compounding household budget pressure? Even a decision about whether to raise interest rates – which may seem easy enough to make based on economics – is not simply a number-crunching exercise. What was the 'right' amount to raise the rate to head off inflation but not induce panic? So that households and businesses could survive, and productivity wouldn't nosedive?

The point I am making is clear: business and economics isn't just a game of hard numbers. Financial decisions are influenced by a complex series of human variables that are, due to the very fact that they involve humans, permeated with uncertainty. At their core, economies are built on human thoughts, emotions and behaviours. Humans create the workforce; are the inventors and innovators, leading to economic progress; and determine the values and standards driving the economy. We can fool ourselves into thinking that uncertainty can be suppressed and controlled by data-based predictive modelling and economic forecasting. As we have seen in relation to education and healthcare, once humans are part of the equation, uncertainty abounds.

Those in business and economics often attempt to minimise and suppress uncertainty – and we can understand why. But in fact their success is often tied not to generating certainty but to how adept they are at managing uncertainty: their level of uncertainty tolerance.

Paying the price of unknowns – how economies, banks and businesses manage uncertainty

In 2015, a report from *Harvard Business Review* revealed that some economies are designed to be less tolerant of uncertainty than others. As an example, the Greek economy was identified by the Geert Hofstede scale as having the highest uncertainty avoidance score (100) out of *all* surveyed countries – meaning this economy actively works to avoid uncertainty at every turn. Similarly, both Portugal and Belgium had high tendencies to avoid economic uncertainty. On the flip side, the United Kingdom, Sweden and Denmark had the lowest scores, meaning these economies are more likely to embrace, or at least acknowledge, economic uncertainty. Countries at both ends of the spectrum appear to build their economic policies around their uncertainty tolerance.

What does this mean, though, in a practical sense? In healthcare, the uncertainty-intolerant system builds rigid structures and procedures to suppress uncertainty and generate a false sense of consistency. It turns out that economies intolerant to uncertainty employ a similar approach. The *Harvard Business Review* article goes on to compare the economic uncertainty avoidance score against the World Bank rating for bureaucracy (that is, the number of rules and regulations governing the banking industry). The levels of economic bureaucracy in Greece, Portugal and Belgium are far greater than in the United Kingdom, Sweden and Denmark. These intolerant economic systems attempt to suppress, reduce and eliminate uncertainty by developing rules, protocols and standardisations – but to what end?

While the sectors of business and economics are not as well studied as healthcare when it comes to managing uncertainty (maybe due in part to a mistaken belief that there is little uncertainty in these fields), the research that has been undertaken reveals a common theme – uncertainty-intolerant systems build workers who are intolerant, and this leads to negative, often unintended consequences.

An area of business and economics where uncertainty tolerance is most studied is auditing. Auditors take a lead role in official reviews of a company's financial and business dealings. This field typically relies heavily on standardised operating procedures and rules to determine whether or not a company is complying with the law. This may convey the impression that uncertainty can be suppressed in auditing – but as we see with healthcare, this is mostly just wishful thinking. At the core, auditing is a human endeavour. Humans create the rules, regulations and procedures carried out and enforced by – you guessed it – humans. Auditors must make judgement calls and decisions about which data matters and the degree of importance of the information provided, and then interpret the data. With every human decision comes a kernel of uncertainty.

A study in 1993 was among the first to shed some light on the relationship between auditors' tolerance for uncertainty and their workplace performance. Auditors' perceptions of autonomy and independence were mapped against their level of uncertainty tolerance. This study found that auditors who were less tolerant of ambiguity appeared more dependent on their co-workers in managing their case loads. Auditors who were more tolerant of uncertainty appeared to need far less team input to make audit processing decisions. As the auditing cases became 'more complex and undefined' (that is, more uncertain), the relationship between uncertainty tolerance and independence became more pronounced, with those less tolerant of uncertainty becoming more and more dependent on others.

Another study conducted in 2004 evaluated auditors' perceptions of risk and time needed to make auditing decisions. This study found that those less tolerant of uncertainty estimated that more auditing hours were required to make an auditing decision than those who were more tolerant of uncertainty. The hours increased for these uncertainty intolerant auditors when the risk involved with the decision-making process was perceived to be higher.

On 12 July 2018, ABC Radio ran a report on the 'big four' accountancy firms – KPMG, Ernst & Young, Deloitte and PricewaterhouseCoopers – and how their work could threaten the global economy. In it, award-winning journalist Richard Brooks highlighted just how frequent, and how important, high-risk auditing decisions are: 'We're all affected by auditing ... we're in a situation where we need really good, strong, effective auditing more than ever.' Auditors ensure that a business's books are accurate and that the business is complying with all necessary legal requirements. We need to trust in these decisions and they need to be ethically sound. When they aren't, Brooks tells us, financial crises can occur.

The results of audits are reported to boards, shareholders and governments; the outcomes can influence investment and policy decisions. If the auditor can't make decisions due to their inability to adaptively respond to the presence of uncertainty, they may not be able to complete the audit, or may become too cautious in their report because the uncertainty confounds them. The knock-on effects of these maladaptive responses to uncertainty can be substantial. The 2004 study showed that experience did not trump tolerance of uncertainty when it came to predicting the risk of the audit or time spent reviewing data in the decision-making process. This means that even senior auditors, who had likely been part of hundreds of auditing decisions and had significant experience with auditing-related uncertainty, still took longer to make decisions than those with *less* experience but *a higher level* of uncertainty tolerance. One could picture a scenario where shareholders are awaiting the data and the data is delayed due to a group of uncertain auditors. In the worst-case scenario, the business being audited could even fail.

The perception of risk and its relationship to uncertainty tolerance is similar in the loans sector. A 2000 study sought to understand how uncertainty tolerance influences decisions to grant or deny loans. The researchers in this study provided loan officers with varying levels of documentation, allowed them to review each scenario and then asked for a lending recommendation. The study results yet again revealed that the key factor in this decision-making process was the level of uncertainty tolerance the employee had. Even when comprehensive documentation was presented to loan officers with a lower tolerance of uncertainty, they still struggled to decide on a loan outcome, letting their uncertainty intolerance cloud the decision-making process.

A study exploring the Greek banking system found another impact of bank administrators' uncertainty tolerance on their work. Here the researchers identified a link between bank managers' perceptions of control in making workplace decisions with their level of uncertainty tolerance. The bank managers in this study were largely intolerant of uncertainty, indicative of the Greek banking system that is equally intolerant of uncertainty. This intolerance led the bank managers to work towards strategies that reduce, minimise and/or attempt to eliminate perceived banking uncertainties. These employees believed that most of the banking outcomes (negative or positive, expected or surprising) were dependent on their own actions in every instance, as opposed to real-world economics, where luck or coincidence also play a role. The system designed to be intolerant of uncertainty created workers who were similarly intolerant of uncertainty. Because of this, these hard-working employees believed that their own actions, and not the uncertainty intrinsic to all banking systems, was the cause of their inability to expect the unexpected.

This hyper-focused, arguably delusional, thinking that uncertainty can be suppressed, be planned for, and overcome in business and economics is having profoundly negative effects. As the expression goes, time is money, and an uncertainty intolerant workforce can cost both. What these studies collectively reveal is that employees who are less tolerant of uncertainty are less able to work independently, take longer to complete their work and rely on others to a greater extent than those with a higher tolerance of uncertainty. This, in turn, leads to increased costs related to greater supervision requirements alongside decreased productivity, not only among a subset of employees but across the workplace as a whole – as those with a lower tolerance for uncertainty often seek advice from those with a higher tolerance. These negative outcomes appear to be exacerbated when the workplace tasks increase in complexity and ambiguity, and when the system within which the employee works is equally intolerant of uncertainty.

Not only are uncertainty-intolerant economies more likely to drown employees in bureaucracy, but they may also cause employee burnout. As employees recognise that the system in which they work rewards certainty, they may begin to believe that they can reduce intrinsic uncertainties. It's this illusion of control – they think they can head off any unexpected outcome if they just gather a little more information, if they just run one more model, if they just consult with one more colleague – that becomes the representative actions in such systems. Despite these efforts to ensure certainty, an unanticipated outcome may still occur – and, more often than not, is *likely* to occur, as business and economics are sectors often marked by volatility and ambiguity. No matter the care, consideration, planning and time put in, the unexpected and unplanned can still occur. Employees in systems that are intolerant of uncertainty, studies such as that of the Greek bankers suggest, are more likely to blame themselves for not expecting the unexpected, and for not preventing it. In addition, instead of accepting that uncertainty is a feature of the system, they are more likely to view unexpected events as able to be controlled in future modelling, even though the nature of unexpected events is that they cannot be foreseen. This disconnect between what such employees perceive as within their control and the reality of economic uncertainty can contribute to overwork and feelings of inadequacy because they are fighting an unwinnable battle – even though they may not realise it.

Interestingly, there is one field of business where uncertainty is embraced and uncertainty tolerance sought after in employees: entrepreneurism. Acknowledging the unknown is the entrepreneur's greatest asset. Entrepreneurs embrace and exploit uncertainty to develop innovations and build wealth. This is evident even in the way they go about planning products: a popular method called 'design thinking' encourages innovators to challenge assumptions and discover the unknown in order to develop innovative solutions.

Entrepreneurs, by the nature of their work, need to focus on novel ideas and developments. Their entire careers are built on imagination. Innovative product solutions occur at the cusp of the unknown. As Henry Ford said: 'If I had listened to my customers, I would have built a faster horse.' Invention is about seeking out uncertainty. Unsurprisingly, a 2010 study found that entrepreneurs displayed a higher tolerance of ambiguity and a higher propensity for risk-taking than small-business owners. The latter value stability over revolution while the former relish uncertainty.

Entrepreneurs can come in many forms, from software developers to app designers to Instagram content producers. Other workers in more traditional business circles may be employed for their ability to display 'entrepreneurial thinking' – to embrace uncertainty and to think laterally in order to come up with solutions. As one participant in my team's work stated:

There are two types of people in marketing. The first are the ones that are more strongly influenced by economics. They're all about certainty. And what they like to focus on is very defined material and strategies that can be learned and regurgitated. It's very difficult to break those people out of that mindset. The problem is that the system rewards that mindset. [This is] as opposed to the other sort of person, which I am, and it's the entrepreneurial marketer type. We look at innovation, we look at change, at new marketing strategies ... the real world doesn't fit in a model. [There's] volatility, uncertainty, complexity and ambiguity, because that's what we're dealing with [in entrepreneurial marketing]. And as marketers, our discipline is all about change. It's not about a static world. And marketing is about interventions and making interventions and complex systems. So you've really got to understand that. And you've got to understand that systems are unpredictable.

My team's research reveals that fields in business and economics that rebel against uncertainty, that actively work to suppress it, share an important factor. As with healthcare, these uncertainty-intolerant fields have, at least in part, a responsibility for people's lives and wellbeing. Economic decisions have the power to impact the livelihoods and standards of living of large amounts of people, whole communities – and this knowledge likely contributes to a decreased tolerance of both risk and uncertainty in the system. This approach, however, contrasts with a swelling of evidence which suggests that simply working to ensure certainty is ineffective. Uncertainty seeps into even the most rigid guidelines, the most standardised procedures. Furthermore, a lack of tolerance for ambiguity creates a less independent workforce, which costs business financially and can lead to staff burnout.

So what is the best way forward? The authors of some studies in this area do offer suggestions for addressing uncertainty intolerance in business and economics. One is to adjust the working environment to meet the needs of workers who are less tolerant of uncertainty. This can be done by offering additional supervision to such employees or building a workplace that focuses on team-focused auditing, for instance.

But these suggestions see uncertainty tolerance as an immutable personality trait, an employee limitation that must be offset. While someone's degree of tolerance for uncertainty was once considered a permanent, static trait, contemporary research, including that from my team, suggests that levels of uncertainty tolerance are changeable with context, influenced by educational and professional experiences. As such, becoming more tolerant of uncertainty is a skill that we can learn, as long as the system we work in is equally tolerant. As one economics professor in my team's study highlights, uncertainty isn't a flaw in the system, it is a feature:

In business and economics, specifically in economics, we think about uncertainty as 1) incomplete information, and 2) unpredictability of outcomes – and uncertainties extend from these two aspects. First, in any situation, you'll never have complete information, despite theory or models assuming all the information is available. Secondly, even if you had complete information, you cannot predict how human beings will react to in any given situation.

What if, instead of adjusting workplaces to support those less tolerant of uncertainty (a process that is likely to slow productivity and cost money), we supported businesses to acknowledge uncertainty, and built systems and structures that help to foster uncertainty tolerance in their workforce?

Making cents of unknowns – how complexity economics embraces uncertainty

The challenge for economics and business sectors is managing the presence of uncertainty while being respectful of risk, given the potential significance of economic decisions on communities. In the balance between risk and uncertainty, the sector often turns to predictive modelling but, as we have seen, such modelling typically relies on past certainties to predict an uncertain future. These models, by design, more often than not fail to predict and prepare us for uncertainties – particularly when we consider an individual family as opposed to a population. The result? Shock when the unexpected occurs and anger at those who failed to warn us as we manage the fallout. But there may be another way.

Business and economics are inherently complex systems. And wherever there is complexity, there is uncertainty. Consider the Covid-19 pandemic example once more. Despite drawing on a variety of datasets from past pandemics, the complex interplay of people's individual choices, businesses' responses, economic leaders' decisions and technological advances rendered many of the early models moot. The disconnect between facts and human-led events and decisions created a chasm between predicted and actual economic outcomes. As the pandemic progressed, economists had a series of revelations: that Covid-19 would spread throughout all regions of the world; that it had a higher disease burden in countries with fewer public health restrictions; and that it could mutate into new, more contagious strains. Most modelling had failed to account for such variables because of an overreliance on historical data to predict current and future behaviour.

A better solution may be complexity economic modelling. This form of modelling creates a space for economic uncertainty.

Complexity economics is the application of complexity science principles to economics. Complexity science is based on the notion that complex systems (systems that function from dynamic, interacting networks) do not function in a linear way. Its foundation is an acceptance and awareness of the intrinsic flux and ambiguity present in complex systems like global economics and business.

I'll illustrate this with a topic I know well: the human body. Metabolism is an example of a complex system and is similar in many ways to economic systems. Nutrients and calories are the body's form of currency. For those seeking to lose weight, we are often told that 'calories in versus calories out' is the way to manage weight. Essentially, this equation assumes that, on the whole, our metabolism is in steady-state equilibrium, and that a simple linear equation can predict a weight-loss outcome. The number of calories we eat minus the number we burn equals and determines our weight loss; essentially, the food energy becomes our bodies economic cost for survival.

Each individual part of the metabolic system is measurable. A food's calorie count can be determined in a controlled environment by measuring energy release following combustion through the aptly named 'bomb calculator'. We also have the power to measure basal metabolic rate (the number of calories needed for your body to function – to take in oxygen, circulate blood and so on) to determine how many calories a person burns throughout the average day. But when we put the data from these individual measurements together, the modelling doesn't replicate the *actual* weight loss of many individuals. Why?

Our bodies are, indeed, excellent at maintaining equilibrium under most conditions, and for this reason the 'calories in, calories out' linear model of metabolism holds up well under *normal* conditions. When we drill down to a personalised metabolic prediction, however, this equation leaves uncertainty unaccounted for. Complex factors, including genetics, activity level, differences in calorie sources, and hormones, can all influence an individual's metabolism. Hour to hour, our metabolism is ramping up, or down, on any given day – it is in a constant state of flux. An unexpected event, such as an illness or a pregnancy, can also change a weight-loss outcome dramatically. The data we put into the modelling is simply an estimation of the average individual's predicted weight loss. But metabolism function, and thus weight loss, are non-linear events. How, when and why an individual burns calories is dependent on a series of dimensional and dynamic communications between our central and peripheral nervous systems, organs and cells, enzymes, and food – it is not as simple as 'calories in, calories out'.

This example illustrates why attempts to simplify a complex system involving non-linear networked events rarely results in pinpoint-accurate predictions. Complex systems don't just have many components; they are comprised of multiple *interacting* components that form a networked complex where each part influences the entire system – not just the next component in an assembly line.

Given this area of economics challenged my tolerance of ambiguity, I followed my own advice by acknowledging this. I sought the insight of a discipline expert who works in an interdisciplinary team to talk me through the value of complexity economic modelling in managing uncertainty in the business and economics sector. I was privileged to learn from Dr Simon Angus, associate professor of economics at Monash University. Professor Angus, as an expert in complexity and data science, is the co-founder of SoDa

Laboratories, which specialises in big and alternative data (data gathered from non-traditional sources). He is also a co-founder and director of KASPR Datahaus, a company that provides customers with real-time global internet insights.

'A car's internal combustion engine has many components – it is very complicated, but it is not complex,' Professor Angus told me. The steps involved in producing the car's forward motion are sequential, repeatable and relatively independent (meaning that each step only impacts on the next step), unlike with the human metabolism, where the energy burned is determined by highly complex networking and communication across organ systems and cells.

'We are okay at average predictions and advice, but we are often wrong-footed when it comes to a particular regime or intervention for a particular person at a particular time,' Professor Angus said. This is because we are not good at employing modelling designed for complex systems.

Traditional predictive modelling assumes equilibrium, a steady state. In this form of economic modelling, data is sourced from the individual components of a multi-component system. Because of this, instead of capturing the interactions *between* the complex system's component networks, this modelling can only predict outcomes uni-dimensionally. As the metabolism example illustrated, drilling down into the components is not necessarily going to build back up to an accurate picture of the complex system – it will, more likely, simply amplify incorrect assumptions in specific contexts.

In contrast, complexity economics assumes non-equilibrium dynamics. Economies are thought to be constantly in flux (non-equilibrium). While traditional modelling, based on a notion of equilibrium or a steady state, is confined to producing the 'typical' or 'average' conditions the model was programmed to assume, complexity economic modelling assumes that if the same complex system is run hundreds of times, there are likely to be just as many different outcomes. Even if we averaged these potential outcomes together, they may not exhibit the kind of mean reversion we're accustomed to; there may be no 'typical' or 'average' outcome. Essentially, complexity economic modelling plans for the unplanned – it models assuming that uncertainty is the likely outcome.

When we review the span of economic history, some aspects of the economy appear stable. For example, Charles I. Jones, a professor of economics at Stanford University, plotted the gross domestic product (GDP) per person in the United States between 1880 and 2008, showing that the ability to predict economic development over more than 100 years is possible, and indeed remarkably accurate. It is the cyclical changes, those *particular* events (such as a pandemic or a war) in a *particular* economy at a *particular* time, that are far less certain. They can bring about unanticipated effects such as shifts in labour and capital, or the emergence of innovative and disruptive technologies. Ongoing evolution, instability and variability is the norm for economies and individual households; economic systems are adaptive and constantly changing. This tenet of complexity economics is not new – it is precisely the view that several economic luminaries, including Thorstein Veblen, Joseph Schumpeter and Friedrich Hayek, held. They too believed that economics is complex and organic, not static and standardised.

Built into complexity economic modelling is the recognition that economic decisions are created, at some level, through the actions of individuals. A complex system has three levels: the individual, the rules of that system, and the interaction between the rules and individuals within a system. A complex system, with its individuals and rules, then interacts with other complex systems. Each system has its own individuals and related behaviours, and its own set of rules. When combined, this new, even more complex system doesn't replicate, at least not directly, any of the components within the separate systems, but rather creates a brand-new complex system with its own unique rules and individual behaviours. Professor Angus tells me, 'Complexity economics is about networks, non-linearity, emergence and self-organisation. The pandemic gave us a good example of a self-organising complex system. Nearly all the ways we worked and lived were changed. Seemingly overnight, economic activity shifted: from family routines; to modes of education and work; to community sport and volunteering; to holidays, leisure and entertainment. The gig economy was there, but it became suddenly enlarged as people were told to stay at home. The speed of adaptation was remarkable. And there was no central controller, no conductor, or even a complicated algorithm, making it happen. Individuals and communities did it themselves. Such adaptation, self-organisation and emergence are key qualities of a self-organised complex system at work.'

Could complexity economic modelling be useful in the example of the Reserve Bank of Australia deciding whether to raise interest rates, I ask. Professor Angus explains its value: 'In the case of the RBA rate hike, a complexity economics approach may engage at the level of the household or even an individual. Rather than making generic assumptions about the median household or average consumer, hundreds of thousands of synthetic households would be created in a numerical simulation. These modelled individual households would be informed by and calibrated to *real* human demographic data. Likewise, thousands of synthetic firms across a range of sectors would be programmed, staffed by the synthetic household members. Economic motivations and needs would be programmed in as well, and the model would be run thousands of times under a range of policy scenarios. The outcome of complexity economic modelling would be a range of realistic trajectories representing potential outcomes for the whole economic system, as opposed to a single output representing an average outcome. In this way, a rate hike's differential impact could be analysed for marginalised and underserved communities, for different types of businesses, or for different societies. The model's capacity to explore the risk of harming one group or another is a key contribution of such a model. This type of modelling takes complexity head-on, complementing existing mean-field or few-type economic modelling.' In summary, complexity modelling provides a range of future possibilities, as opposed to a single average possibility.

Work in the 1920s and 1930s by economists Frank Knight and John Maynard Keynes began to unpack the differences in the relationship between risk and uncertainty. With risk, you can calculate uncertainties; but with unknowns, calculating probabilities will not lead to certainty. We can calculate the likely impact that a rate hike could have on household spending, for instance, but we don't have an equation to predict what the action of those who *don't* adhere to this modelling will be – we don't have a probability equation for the null, essentially. The variables of how, to what extent, and in which ways someone would not adhere to the modelling are too diverse. Instead, complexity economic modelling is programmed to include multiple possible individual actions, providing an output inclusive of this potential 'uncertainty'.

Balancing the books – building uncertainty tolerance in business and economics

Economies that embrace rather than suppress uncertainty in their economic modelling are more likely to be resilient because of their enhanced capacity to prepare for the unexpected. Instead of focusing on the expected average outcome, these economies can gear up for multiple, complex, dynamic and uncertain outcomes. Economic policymakers may achieve the best result by employing *both* traditional predictive and complexity economic modelling. This approach (encompassing both equilibrium and non-equilibrium dynamics) can arguably lead to fewer unpredictable events, as it aims to identify both the likely, or 'typical', outcomes, and a range of less-probable consequences.

An example of what this could look like is represented in the dynamics of the classic iterative logistic map, depicted in the figure below. Professor Angus describes this image: 'In this simple iterative system, inspired by population dynamics, phases of stability (solid lines) sit alongside phases of chaos (diffuse dotted sections). In the former, the system settles to a highly predictable pattern. In the latter, the system exhibits thousands of potential states, confounding simple or certain characterisation.'

We could imagine that traditional predictive modelling would likely predict the phases of stability, the solid lines. Predicting the less certain but still possible outcomes of the unstable-phase model, the diffuse dotted sections, is where complexity economics approaches would excel. In this way, complexity economics can help us to *see* the uncertainty. Once we are aware of its existence, we can develop a plan for addressing it. Essentially, complexity economic modelling orients us towards uncertainty.

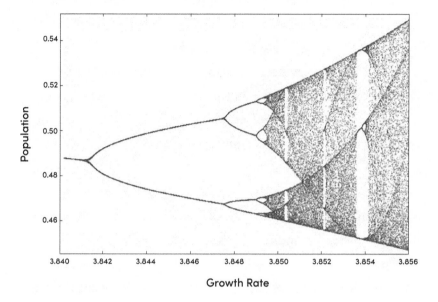

This figure illustrates how complex systems, in this case population growth, can develop regular and predictable patterns when viewed as an overall trend, such as the fractal pattern seen here. When we look closely at a single location point along the x-axis (growth rate), numerous possibilities become visible. Similar to a photo mosaic, there is the appearance of certainty at a distance, but when you see things up close, the diversity and uncertainty are far greater.

When it comes to economic policy, just as we saw in healthcare, the best way to manage uncertainty is not through universally applied rules and regulations. Instead, economic policies need to be applied flexibly to a given context at a given time. By building complexity economic modelling into economic policy development, we can help develop economic systems that are less averse to uncertainty and more resilient to inevitable economic volatility. We can also develop economic policies that do not just satisfy the 'average' person but are adaptable to those most in need.

By building economic modelling to take in diverse data, including that related to the underserved and often unseen, we can create economic policies that help and support more people. Central to all of this, however, is an increased awareness and acceptance that uncertainty is part of the system, not a fatal flaw.

Small-business owners can also prepare for inevitable uncertainties. Creating a work environment where seeing uncertainty is encouraged is a great place to start. In addition to recommendations outlined in Chapter 5, workplaces can build into meetings opportunities for individuals across the business to offer their opinions and suggestions about spending habits. Many employees will know the business well and have insight into the nuanced spending habits and relevant considerations of why, when and how money is spent. Identifying a business's values and priorities, and ensuring that these factors are considered in the financial modelling, is also important. Instead of viewing uncertainty solely as risk, we can build in opportunities to manage uncertainty through flexible standards. Just as we saw with healthcare, if the business standards and practices are too rigid, they are likely to break instead of bend in the face of uncertainty. Instead of disengaging with the uncertainty that a disrupter brings to the sector, a company can explore the possibilities this gives rise to.

How do you do this? One approach is to take a leaf from design thinking, which foregrounds the consumers' needs. We could co-design the future of small businesses with owners, patrons and employees, using the principles of design thinking. A 2015 paper authored by Professor Pavie Xavier and Dr Daphné Carthy, from the École Supérieure des Sciences Economiques et Commerciales Business School, outlines how you can 'leverage uncertainty' through design thinking to focus on innovation – a necessary step for any small-business owner facing a sector disruption or an unanticipated event. The key point in this paper is how this approach supports effective uncertainty tolerance in a responsible way – something that many view Amazon as *not* doing.

Xavier and Carthy underscore a key idea of this chapter, that 'uncertainty is not exclusive to technological innovations, but rather applies to all profitable implementation of new ideas and their corresponding process': 'The action of progressing through the latter can be described as muddling through, where each step is taken into the unknown. Indeed, any innovation process, by definition, requires action to be taken under conditions of uncertainty.' The author team describe design thinking's three-pronged approach to tackling complex, uncertain ('wicked') problems: 'desirability (human needs); viability (business needs) and feasibility (technical needs)'. To do this, there are five key steps. 'Understand', the first step, focuses on deeply exploring where the problem sits. This is where the problem statement is refined to its core. In the case of a bookshop addressing the threat posed by Amazon, this might include aspects of earnings loss, ease of access or challenges related to availability. Next, the group engages in co-creation. Here, the sky is the limit and half-baked ideas for addressing the problem are encouraged. It isn't until the co-evaluate stage that these limitless ideas are given boundaries, where feasibility is considered, and where discussion focuses us towards the next steps – development and implementation.

Diverse problem-solving teams shine in the face of uncertainty. Just as we see with diverse teamwork in education, the variable experiences and perceptions of the co-design team can shed light and perspectives that a single business owner alone may not. We will explore other ways of applying such approaches to 'wicked problems' in Chapter 6.

While we can use design thinking for managing uncertainty in business, we may be able to draw on the lessons of complexity economics in managing our personal finances. Instead of modelling our household budgets solely on our average spending over a historical time period, what if we modelled our budgets from a series of different people's household data? I could see a future where a budgeting software program amasses data related to spending behaviours across all users and creates a personalised budget from relevant datasets. The de-identified budgeting data collected from all users could also be used to build a collective dataset for relevant 'unexpected' events (e.g. employment change, marriage, major holiday, illness, interstate move and so on) to help predict actual spending behaviour during times of great uncertainty. The user would ask the budgeting tool for scenarios (by ticking boxes for the scenarios they are interested in seeing, e.g. marriage, a child) and the algorithm could run through thousands of potential future financial outcomes unique to their individual situation, representing a household form of complexity economic modelling. Perhaps the software could even produce a map of potential outcomes – a bit like the population dynamics diagram – that allowed users to see the likelihood of each outcome.

If budgeting tools became very sophisticated, the software could even give recommendations to manage different potential outcomes (for example, buying more insurance, taking out a loan, putting money in savings and so on). Each user would have the power to decide which of these economic outcomes to plan for – because they deem the degree of risk too high not to – and which to put to the side and deem as lower risk.

While this complexity economics budgeting tool is entirely imagined, the approach to modelling is available to us now. Instead of creating a budget based on an average of your last three months of spending, why not try running through a series of budgetary predictions that incorporate different scenarios (such as job loss, illness, relocation, interest rate hikes, energy cost changes and so on) to try to capture and plan for the uncertainty that traditional budgets fail to encompass? A word of warning, though – to get the true benefit from this exercise, you may need to first develop your own uncertainty tolerance. Without this, the many 'uncertain' financial outcomes could be confronting and/or fear-inducing.

Uncertainty-tolerant economic and business sectors depend on a workforce with the capacity to effectively manage uncertainty. What if we could foster a greater tolerance of uncertainty among economic policymakers, business owners and employees? Given the repeated finding that experience alone doesn't create uncertainty tolerance, it is essential that we develop an educational system that is purpose-built to foster uncertainty tolerance in learners before they enter the workforce. This solution was also proposed in the *Harvard Business Review* article about uncertainty-avoidant economies.

My team's research revealed that many educators who focus on preparing business and economics students for the multiple sources of uncertainty they will face in their careers use a shared stimulus of uncertainty, drawing on real-world data or industry-based placements to support their teaching. One approach to introduce students to uncertainty early in their degree could involve inviting students to voice their preconceptions of their discipline (for example, 'numbers are facts, so accounting is always certain') and then presenting a workforce panel whose members describe how prevalent uncertainty actually is in their careers. Once students are primed with an awareness that learning to manage uncertainty is crucial to their future success (and thus have a sense of purpose in navigating uncertainty in the classroom), the students can be challenged to enter the 'tunnel of uncertainty'.

Each year, one of our study participants, a professor of economics, picks a different country's economy for the class to explore. Students are tasked to answer three questions related to this country's economy: what are the main economic challenges facing this country? What are the possible sources of these challenges? What should the economic policy response be to address these challenges? Real, undigested, incomplete and messy economic data from the selected country is provided to the students (a critical incident uncertainty stimulus). In diverse teams, the students work together to determine their answers to the three questions, culminating in a final group presentation to the class. The final presentation reveals that, even though all teams explored the same country, the answers to the three questions are diverse. Groups identify different economic challenges and different potential solutions. Usually, many of their proposed solutions are viable. Lastly, the professor gives a summary of the actual economic response the country took, and students have one final challenge: they must decide whether the country's course of action was appropriate and defend this position to others in the class.

This approach to teaching uncertainty tolerance – by providing students with a case-based critical incident, as discussed in Chapter 1 – requires students to rely on the principles and knowledge of economics, while illustrating to them that these models and theories bring with them elements of uncertainty. Students report that this approach helps them to recognise the divide between theory and practice (and how to integrate the two) that they will carry forward into the future. This helps the students to develop a skillset for navigating uncertainty in their future careers.

This professor's approach is not unique. There are multiple reports of case-based inquiry supporting learners' uncertainty tolerance in business and economics. Kevin C. Banning, a professor of management at Auburn University in Alabama, reports similar learner impacts in a 2003 study, underscoring the positive influence that educational approaches can have on helping students to develop this essential workplace skill. In Banning's study, as in our participant's teaching, how the critical incidents are facilitated (to highlight and underscore the uncertainty, complexity and ambiguities) is a key factor in their success.

For more advanced business and economics students, simulated or immersive work experiences are a valuable way to help prepare them for future workplace uncertainties. One participant in my team's research described a 'marketing challenge', where local companies pitched a problem to class teams, who had to work together to solve it: 'We had a client, we drew on the client's problem, we solved the client's problem and sold our solution. So instead of dealing with a small bit of [uncertainty], we were dealing with the real-world uncertainty. Unless you've been in [the] industry and realise that the world is not a predictable, comfortable place, and you've got to deal with crazy things, you don't experience the full spectrum of [the] industry.' Here, the client's problem isn't the only source of uncertainty. In these destabilising and iso-immersion educational practices, the sources of uncertainty extend to the client's perceptions and goals, the team dynamics, the consumer's feedback and so on.

If, instead of seeking to suppress and control uncertainty, we build the unknown futures into our modelling and help prepare the forthcoming workforce for it, we are more likely, both as a society and as individuals, to approach the inevitable uncertainties of economics and business with solace and reassurance, with hope and curiosity, instead of anger and alarm. We are more likely to accept that sometimes, no matter how well we plan, uncertainty will derail our business dealings and our finances – but we will also be more capable of feeling better equipped to deal with the effects of this.

Cashing in on uncertainty – managing future economic challenges

Among the greatest challenges facing economies today is climate change. Predictive economic modelling on the impacts of climate change has the potential to mirror the unpredictability of early Covid-19 pandemic economic modelling. Both of these events were unprecedented. Just as with the pandemic, we do not have large historical datasets to feed into the modelling to represent the 'typical' impact of climate change on our economies. We can plug in bits and pieces of related data from current climate events, as we saw economists do in the pandemic context with some of the models used to predict the economic fallout from Covid-19. But just as then, when we reduce complex systems to their parts, and try to draw these pieces together to make predictions, the capacity to model real-life, multifaceted, dynamic systems is diminished, if not entirely eliminated.

Complexity economics, however, may allow us to model for the economic uncertainty that accompanies climate change. A complexity-economics perspective would consider animal and plant behaviours as well as human behaviours; planetary health rules as well as human economic rules; and how different ecosystems interact across the globe, along with how different national economies interact globally. Instead of viewing the interactions across these three domains as sequential and linear, complexity economics assumes dynamic, multidimensional interactions. I can't take credit for this idea: in 2015, a key figure in complexity economics, Professor J. Doyne Farmer – a leader of the Institute for New Economic Thinking at the University of Oxford – published a paper titled 'A Third Wave in the Economics of Climate Change', highlighting the significant role of complexity economics in predicting the 'economic impact of the physical consequences of climate change'.

In planning for the impacts of climate change on economies, ecologists and earth scientists will need to work alongside economists to address the challenges facing the globe. This interdisciplinary team will need to build modelling scenarios that consider a range of human choices alongside impacts on, and responses by, the plant and animal kingdoms. This team will need to be able collaborate across disciplines, sectors and fields. Let's hope their educational and workplace experiences have prepared them for that.

Many world leaders are, perhaps inevitably, intolerant of the economic uncertainty that climate change brings. In response, some global leaders choose to focus on the knowns of today, instead of the unknowns of the future. But in prioritising familiar, traditional predictive economic models, global leaders are not encouraging us to plan for the inevitable uncertainties around planetary health. Indeed, an intolerance of uncertainty, specifically about the benefits and risks of 'going green' with economic policies, may be what is hindering some countries from taking action now to minimise the adverse climate impacts in the future.

The sum total

In economics and business – sectors so heavily focused on numbers – uncertainty can feel anathema. But uncertainty will exist regardless of our attitude towards it. And complexity is something that the human intellect is uniquely primed to navigate. Humans outperform AI as tasks get more complex and uncertain – AI's abysmal performance in picking 'winning' stocks in the stock market is just one example of this. No matter how technologically advanced these fields may seem, or the extent to which AI is playing a supportive role, ultimately humans are the ones making decisions – and this is a good thing.

Global economies, businesses and individuals are faced with a choice – embrace and plan for the inevitabilities of uncertainty, or risk psychological distress, increased work burdens and failed predictions. While business and economic leaders may attempt to eliminate uncertainty with predictive modelling, rules and regulations, and inflexible policies, these strategies do not stave off the storm; they simply lull us into a false sense of certainty. As individuals, we should challenge traditional predictive modelling before accepting the predictions wholesale. We should learn to ask about the data sources for the modelling and explore the types of scenarios the modelling was programmed to undertake, and seek to understand what doesn't quite fit.

If we embrace uncertainty, model for it and better prepare the future workforce to navigate it, we can create economies and businesses that are resilient in periods of economic upheaval. By encouraging complexity economics modelling and interdisciplinary economic solutions, and embedding uncertainty-tolerance teaching practices in business and economics education, we can move towards solving tomorrow's problems today.

4

UNRAVELLING LIFE'S MYSTERIES COLLABORATIVELY

How uncertainty affects scientific discovery

'It ain't what you don't know that gets you into trouble. It's what you know for sure that just ain't so.' —MARK TWAIN

A seasoned and accomplished researcher who participated in one of my team's studies once said to me, 'The whole point of science is not knowing and finding out. That's the exciting bit.' From Nobel laureates to early career researchers, there seems to be a particular enthusiasm for the unknown among those who make their livelihood as scientists. Researchers often consciously choose a career that has uncertainty at its very heart. Science seems to begin and end with uncertainty.

The retelling of scientific discoveries acknowledges, in part, this relationship between science and uncertainty by labelling many results as 'happy accidents'. For instance, penicillin was discovered only after Dr Alexander Fleming *accidentally* left out his experiment to grow mould; Viagra was originally used for chest pain, but Pfizer *inadvertently* noticed that erections were a side effect in its clinical trials; the microwave came about when Percy Spencer *unintentionally* realised that the chocolate bar in his pocket melted when he went near the magnetron ... the list goes on. We romanticise scientific discovery, which comes from creativity, exploration and mystery – all trademarks of uncertainty. As American theoretical physicist Richard Feynman put it: 'It is absolutely necessary, for progress of science, to have uncertainty as a fundamental part of your inner nature.' Our society seems to relish uncertainty in science – only when the uncertainty is in the rear-view mirror.

Uncertainty is a much bigger, broader and more powerful force in science than our neat retellings of scientific discovery illustrate. While we may laud scientific discoveries that happen by accident, when it comes to scientific research, both the public and the scientific community seem to view uncertainty through a deficit lens – something that should be reduced, eliminated or suppressed. Documentaries seem to narrate scientific breakthroughs as a set of relatively linear and sequential steps that progress towards addressing and closing off uncertainty, and rarely address the impacts of human sources of uncertainty on this journey to knowledge – humans introduce uncertainties, often unknowingly or unconsciously, into scientific research. We think of uncertainty as a shortcoming of research, as opposed to a fundamental feature.

In the pursuit of uncertainty

Let's consider the stories told about the 'accidental discovery' of penicillin – a life-changing antibiotic that has shaped the course of medical treatment for generations. Search for the history and you will find articles with titles such as 'The Fairytale

Discovery of Penicillin', 'The Forgetful Father of Penicillin' and 'Alexander Fleming's "Eureka" Moment'.

We typically hear tales of scientific discovery built around a scientist-hero who 'sees' things that others don't. These heroes tend to be archetypally male, white and English-speaking. But in practice, science is not a field of singular events which lead to miraculous findings due to sparks of insight from individuals. It is a field of collaboration, of regular failures, of human choice and autonomy, and of convoluted paths to discovery. With the power of hindsight, we craft these fables that whitewash the messiness. We trade the discomfort of uncertainty for scientific legends and palatable certainty.

The main character in the popular telling of the discovery of penicillin is an archetypal scientist, Dr Alexander Fleming, a noted researcher in the field of bacteria. He is about to leave for a holiday, and in his rush for a break, he forgets that he had left some prepared petri dishes out on the bench.

Upon returning from his holiday, Dr Fleming, the ever-inquisitive scientist, notices a mould growing on the petri dishes he left out and realises that no bacteria are growing near the mould, despite these petri dishes being designed to support bacterial growth. Dr Fleming wonders why the bacteria aren't growing, and whether the mould had something to do with it. In these tellings, Dr Fleming's superhero power includes his keen observation – unlike others in the field, he was the first to question this anomaly, if not the first to see it. With perseverance, Dr Fleming finally identifies a substance produced by the mould that prevents bacterial growth, and he names this incredible anti-bacterial substance 'penicillin'.

This is typically where the story ends. Dr Fleming is the hero in a story for seeing what others didn't and pursuing this observation to its natural conclusion – the discovery. There is no mention of failures, no mention of the mishaps or other people involved. After all, who would be interested in that?

But the discovery of penicillin doesn't end here. Discovering the antibacterial substance on its own is pretty useless in terms of medicinal effects. Knowing that the substance exists is only the first step – isolating and making the antibiotic at large scale is critically important for penicillin to have value to human health. When Dr Fleming tried to isolate the penicillin from the mould, he failed – repeatedly. He sought help from expert after expert, from pharmacists to chemists and everyone in between, but he continued to be unsuccessful in isolating the substance.

He resigned himself to using the mould with the penicillin, instead of isolating the penicillin. By using the 'mould broth' he was able to show that the broth (as a whole) had medicinal effects on the wounded. The broth seemed to kill off infections, but only when the broth was applied directly – meaning the scientific community couldn't be sure that penicillin was the substance leading to the bacterial death. Nor did they know *how* this substance was working.

The Science Museum in London reports that Dr Fleming declared that 'the production of penicillin for therapeutic purposes ... [is] almost impossible'. Despite being the sole winner of the Nobel Prize for Science for his discovery of penicillin, he wasn't the one who isolated it, and it appears that he distrusted that penicillin isolation was even possible.

There are reports that Dr Fleming's penicillin 'discovery' was soon dismissed by the scientific community. Not only was he having trouble isolating it from the mould, but scientists at the time were hesitant to embrace the idea that toxic substances (including mould) could be useful to treat human illness. This dogma added to the growing uncertainty about the future of penicillin. The idea that a substance from a mould would be valuable to healthcare, and taken as a treatment, didn't come for years – and wasn't led by Dr Fleming. To quote German physicist Max Planck: 'A new scientific truth does not triumph by convincing its opponents and making them see the light, but rather because its opponents eventually die, and a new generation grows up that is familiar with it.' This quote underscores how culture and the paradigms of the day introduce human uncertainty into the scientific process, influencing the way results are interpreted, perceived and pursued.

Eventually, the scientific community became more open to the idea that ingesting potentially toxic substances could have therapeutic effects and Dr Fleming's research was met with renewed interest. The story of penicillin now continues a decade after Dr Fleming is credited with its discovery. A team of researchers led by Howard Florey and Ernst Chain uncovered Dr Fleming's old research notes – and thought that there was something worth pursuing further.

The Australian writer and medical doctor Peter Goldsworthy and University of Adelaide professor of psychiatry Alexander McFarlane describe Howard Florey as 'an abrasive Australian ... who was methodical to the point of obsession'. Ernst Chain was an expert in lysozymes (cellular machinery with enzymes that can digest bacteria). This scientific duo, with their complementary work styles and expertise, focused on research that addressed the antibiotic resistance increasingly present in patients. Their curiosity about the potential of penicillin drove them to persevere through years of failed attempts at isolating it. World War II was raging at this time, and bacterial infections from battle wounds were killing soldiers. This, alongside the changing views of the scientific community, drove funding, interest and support for Chain and Florey to continue with their work. From this, we can begin to appreciate that the context within which research is undertaken often influences what is studied and how it is studied – generating additional potential sources of uncertainty that may impact on the process of discovery.

There are some reports that the team of researchers working with Chain and Florey were often found in the throes of argument as they attempted to figure out a way through the uncertainty related to penicillin isolation. Eventually, with painstaking commitment, after multiple failures and many, many hours, the team was able to develop an inefficient way of isolating penicillin. The procedure was adequate enough to test some isolated penicillin in clinical trials. The results were sufficient to support continued investigations to improve the production and development of this life-saving medicine.

Goldsworthy and McFarlane highlight in their *Medical Journal of Australia* article that Florey 'shunned the media for fear of creating false expectations and had nothing but contempt for their intrusions into his life'. While many scientists in contemporary contexts are beginning to engage with the media, publishing articles about their work for general audiences, there remains a hesitancy from some scientists and researchers to engage in public discourse because, similar to Florey, they are concerned about misrepresenting their work.

Goldsworthy and McFarlane also highlight societies' strong desire to showcase scientific discoveries as 'quests'. These plots typically have a main character who sets out on a challenging journey to find a notable artefact. The hero in our story, Dr Fleming, faces the obstacle of the scientific communities' disbelief in the value of this potentially life-saving artefact. Against all odds, he perseveres and fights against the conventional logic of the day. Ultimately, his discovery leads to the saving of countless lives.

In reality, the journey to penicillin had multiple leading characters, was influenced by the societal paradigms and challenges of the day, and took decades to complete. Yet we rarely hear about all the mistakes and wrong turns that were actually necessary steps towards discovering a life-saving medicine.

Our society is so enamoured with these quest narratives that an entire genre is devoted to them, with such titles as *Serendipity: Accidental Discoveries in Science* and *The Amazing Accidental Discoveries of 100 Brilliant Scientists*. The subtitle of *Accidental Medical Discoveries* is 'How Tenacity and Pure Dumb Luck Changed the World', as if science is a mixture of heroism and magic that ends in certainty. Few in the scientific community or the media communicate the complex and uncertain reality of research activities. The waiting, the reading, the mistakes, the wrong turns, the complexity, the prolonged and collaborative nature of investigations – rarely are these parts of the documentation of scientific discovery. Instead, we tend to hide the scientific uncertainty in the telling of a quest journey that ends with certainty – a discovery wrapped in a bow.

Why are we so wedded to telling stories of scientific discovery in this way? In part, because the system supporting scientific discoveries appears to actively suppress uncertainty by limiting avenues to share these realities. There aren't journals that support tales of the failures, perseverance, frustrations and anger – the uncertainty of science. The Nobel Prize for Science isn't awarded for research reports about boredom and waiting, and is rarely awarded to more than one person (let alone to multidisciplinary teams). Grant funding is finite and seldom supports the decades typically required before the end point. The paradox of science is that research is a journey into uncertainty, but it is certainty that tends to be rewarded.

The supporting characters of scientific discovery

This hero trope of scientific discovery, and the concealment of the drawn-out process and many individuals involved in scientific research, has real, often negative, consequences. Not only does this approach ignore the uncertainty at the core of scientific revelation, but it disregards the breadth of human involvement required to progress towards new knowledge. Teams of researchers, or 'supporting characters' in the hero scientist's story, are often left out entirely, and those individuals erased from the discovery quest tend to also be those in traditionally marginalised communities.

Think of the quest tale about the discovery of DNA structure. The heroes in this story are the famed James Watson and Francis Crick. We may now be more familiar with someone who was left out: Rosalind Franklin. Dr Franklin's expertise in x-ray crystallography led to images that helped reveal the double-helix structure for which Crick, Watson and Maurice Wilkins share the Nobel Prize for Science. Yet she died – at thirty-seven, of ovarian cancer, which some believe may be linked to her exposure to radiation – with her work largely unrecognised by the scientific community. As a 2008 *Nature Education* paper put it aptly: 'The landmark ideas of Watson and Crick relied heavily on the work of other scientists. What did the duo actually discover?'

Even as I write this book I feel the discomfort of claiming any work related to uncertainty tolerance as truly my own.

As a researcher, I am acutely aware of the critical impact that the collective body of knowledge across this field has on my team's understanding of this concept. Indeed, in my own research experience, it is the teamwork that leads to more thorough understanding and more well-rounded discoveries, and is the source of joy through the uncertainty.

In 2022, a top-tier scientific journal published the study 'Women are Credited Less in Science than are Men', which underscores the negative consequences of these quest narratives with archetypal heroes. The reason behind this lack of credit? Women are less likely to be named authors on papers and patents. When women are named, their roles are downplayed or their work becomes 'invisible', unappreciated or 'ignored' past the co-authorship – and this omission of women seems to happen at every stage of their careers. Even the more senior woman are hushed in the communication of science. These types of omissions have big implications, not only on those impacted by this omission but on the field of science itself. A recent paper in the *Proceedings of the National Academy of Sciences of the United States of America* (PNAS) led by Dr Yang Yang, an assistant professor of IT, Analytics and Operations at the University of Notre Dame, Indiana, explored the impact of gender diversity in research teams and found that 'mixed-gendered teams – teams combining woman and men scientists – produce more novel and more highly cited papers than all-woman or all-men teams'. By not supporting diversity in research teams, we may actually be hampering scientific discovery.

These invisible contributions to science are becoming more visible in popular media, giving the public an opportunity to learn a bit about the how and why of these omissions. The 2016 film

Hidden Figures, for example, focused on the contribution three African-American women made to the success of NASA's flight missions, and how racism and misogyny impacted on their public recognition for this.

Sadly, though, there are likely far more 'cover-ups' in the field. Another example? Alice Augusta Ball, a pharmaceutical chemist who discovered a pain-management treatment for leprosy. According to a *National Geographic* article published in 2018, Ball was 'the first woman and first African-American to earn a master's degree in chemistry from the College of Hawaii. After her untimely death, Dr Ball's discovery was taken up by the then college president, who failed to credit her for the discovery.' In essence, we haven't come far in our acceptance of diversity in science since the 1940s. In Chapter 5 we will see how diversity and inclusion is linked to uncertainty tolerance, and this lack of acknowledgement of diversity is yet another mechanism for suppressing uncertainty in science.

These stories might excel at suppressing the discomfort of uncertainty, but who 'made' the discovery if we acknowledge the teams and decades of time? As four senior academics – Melissa Nobles, chancellor and professor of political science at Massachusetts Institute of Technology; Dr Chad Womack, senior director of national STEM programs and initiatives at United Negro College Fund; Professor Ambroise Wonkam, director of McKusick-Nathans Institute and Department of Genetic Medicine at Johns Hopkins; and Elizabeth Wathuti, climate activist and founder of Green Generation Initiative – note in a piece for *Nature* in June 2022, by questioning the racist and gendered tellings of scientific discoveries, we also end up questioning the foundations of science, including the evidence that underpins it and the methods used to discover it. Thus, to make science more equitable, less biased and more inclusive, we must face and seek the uncertainty in the scientific process. Instead, for far too long we have clung to digestible quest stories that reinforce a biased view of what an expert looks like (a white male scientist in a lab coat who has the superpower to see what others don't) – but at what cost?

The cost may be a miscommunication of the extent of uncertainty within the scientific process, including the uncertainty introduced by working in diverse teams, and the value this uncertainty has in helping us learn and discover. These stories elevate only certain 'expert' opinions, leading to a misconception about whose opinion is considered valid in the wider community, silencing critical contributions to scientific discourse and teaching us to value the scientific knowns over unknowns (the end result over the convoluted process that leads us there).

An experimental paradox

From the outside, there appears to be an experimental paradox – science is grounded in uncertainty but is engaged in the pursuit of truth and certainty. Driving research initiatives are humans (or teams of humans), who use their intellect and autonomy to make decisions about the way the research progresses, the outcomes they seek to identify and how the findings are interpreted. The paradox arises from the disconnect between our attempts to seek objective certainty and the natural uncertainties present in the complex system we seek to study.

The Covid-19 pandemic is an ideal representation of the exponential sources of uncertainty in science. The research subject (the novel virus) was unknown. What we knew at the beginning of the pandemic was far outweighed by what we *didn't* know. We knew that the pathogen was a virus, and which type (a coronavirus, in the same family as the common cold). But, at this early time point, we couldn't even fathom all that we didn't know. These unknown unknowns, or blind spots, were enormous.

In scientific research there are also known unknowns (identifiable gaps in knowledge). These are questions researchers know to ask on topics we don't yet understand. These were plentiful with the onset of the pandemic. What made people susceptible? How was it spread? What were the short-term and long-term consequences of Covid-19 infection? What was the mechanism of the viral entry and reproduction? Were there any existing treatments that could help protect people? At what point should we call this a pandemic? These known unknowns appeared endless.

At the start of the pandemic, when the unknown unknowns were greatest, scientists were at the threshold of a series of tunnels of uncertainty, so many they felt they were navigating a dark labyrinth – thresholds were present at every turn. Researchers made a choice – in record numbers, according to the research paper outputs – to stand at the mouth of the labyrinth and walk in.

During the pandemic, there was a divide between the way scientists understood this labyrinth and the way some in the public interpreted it. Most scientists knew that this labyrinth wasn't a static structure. They knew they were running an 'amazing race' of research, with the known unknowns of viral mutations constantly changing the contours of the labyrinth. Researchers knew that the virus would likely mutate, but they didn't know how fast or what the impact of each mutation would be. So, with each piece of new knowledge, the labyrinth moved around them – the walls shifted, and some routes that they were following closed, while new ones opened. The shifting parameters of scientific discovery aren't unique to Covid-19. The world we live in is changing and evolving at all times, and what was 'truth' may, as time marches on, no longer apply.

This is where the paradox, or rather pseudo-paradox, of research comes in. Researchers recognise that scientific knowledge is never certain, because what we know applies to the context of what, when and how it was studied. If any aspect changes or shifts over time, what we 'knew' may no longer apply.

For instance, what we came to know about Covid-19 early on didn't necessarily apply in later stages of the pandemic, because the virus changed, our knowledge about it changed and we changed. While surgical masks and social distancing may have worked earlier in the pandemic when the original strain was circulating, N95 masks and improved air filtration became the gold standard as the virus mutated. Some took this to mean that we 'don't know anything' or that 'the science was incorrect' – but this is a fallacy. Just because the context no longer fits the original findings, it doesn't mean that what we discovered was wrong. Science is never done 'knowing' because the complex system we study is ever-changing.

Researchers move from *unknown unknowns* to *known unknowns* to *known knowns* as through a labyrinth. Scientists at a start of a research project have some idea of the known unknowns, so can identify some entrances into the unknown (panel 1). As researchers across the globe work on their individual projects, they combine their findings with what others are learning to put together a picture of the known knowns (panel 2). Once the known knowns become plentiful, scientists can follow the lightened path towards 'settled science', or science that is relatively stable no matter the context. But scientific knowledge is always evolving. For every lit labyrinth, there is another waiting to be explored (panel 3).

As researchers found their way through the Covid-19 labyrinth, the virus changed, evolved and spread. Researchers had to decide which way to turn, with some directions leading to dead ends. The uncertainty wasn't only about the lack of knowledge of what was being studied, but also about how to interpret the available findings, and whether these findings would apply tomorrow or in three months – and whether they were applicable in another context or culture.

As scientists navigated the labyrinth, they sent out flares by sharing their research findings through scientific journals, social media and news outlets. To those of us on the outside, who weren't seeing the day-to-day journey through a dark maze, the flares could seem chaotic, haphazard and random. But to the scientists, the flares were indications that the research was working as it was meant to. They gave a signal to others in the darkness, who could adjust their path accordingly. As the data accumulated, researchers begin to discreetly identify the regions of the labyrinth they were familiar with and section off the areas that were still unknown. The more scientists were able to label, describe or identify the unknowns, the more scientists could move into the area being studied and close in on some answers about it. In this way, scientific proof is built over time and through collaboration. But such 'truths' will always carry a level of uncertainty, because scientific experts anticipate and expect that the knowledge may not apply forever, or to all related situations.

For example, when the meteorologist describes a path of a cyclone, they illustrate both the 'likely path' and the 'potential paths'. When describing the potential path, the weather person typically includes a description of what factors might take the cyclone path in one of these alternate directions, recognising the impacts of a complex and changing system on the 'truth' – they are acknowledging that parts of science remain 'unknowable with precision', as complexity scientist Geoff Boeing states.

All science is based on some known knowns. These are fundamental processes in science that have stood the test of time, are relatively stable across contexts, and serve to explain many situations accurately. In the weather example, the known knowns include how cyclones, in general, form and behave. Because we know so much about this, we can then identify the known unknown: how will *this* cyclone behave? So just because we can't predict the exact pathway of this exact cyclone on this exact day, doesn't mean we don't know anything about it, and that science can't be trusted – it means that what we know has degrees of uncertainty related to it, and we acknowledge this. It is because we know so much that we are even able to identify the general path of the cyclone, as well as its potential paths. But this pseudo-paradox – that when science is working it has degrees of uncertainty, and that this uncertainty is what leads to scientific evidence and proof – is what can generate distrust in science, and it is often exploited.

An article in *The Conversation* from July 2022 highlights just how impactful this pseudo-paradox is. A trio of experts in psychology and human behaviour – doctors Aviva Philipp-Muller,

Richard Petty and Spike Lee – outline reasons why people typically reject science. The top reason is that they question scientists' credibility. What causes people to question scientists' credibility? The changeability of science! The very nature of offering advice on a complex world leads to distrust. But why do so many people distrust this variability instead of rejoicing that science is working as intended?

One of the solutions posed to address this distrust is to be more transparent about the omnipresence of uncertainty in science research. Yet our scientific community seems to struggle with being open about the uncertainty inherent in science. While scientists may thrive on uncertainty and relish the challenges they face in their research, the desire to reflect on and share these uncertainties is limited by the system within which they work.

To understand the solutions to the pseudo-paradox of science, and the related distrust, we must explore the sources of uncertainty in the research method more closely.

Uncertainty under a microscope

There are two key sources of uncertainty that scientists contend with in research: the uncertainties of the complex system being studied, and the uncertainties related to the humans undertaking and driving this research.

Let's say, for instance, a researcher wants to understand how different molecules are transported across a cell. The transport pathways and cellular machinery don't work as a conveyer belt or assembly line. There are external and internal environments (pH, temperature, resources) that impact on this machinery to produce a series of interrelated events – all of which affect the original pathway of interest. The type of molecule being transported can also influence this cellular system. Cells also work within complex tissues, multifaceted organ systems and the intricate human body. Human researchers interact with another complex, global system made of diverse cultures. As we can imagine, attempts to study a single pathway in a cell is not as straightforward as we would all like it to be.

Not every research study can consider these levels of complexity, and many studies aren't designed to – they are designed to focus on the topic of interest through purposeful exclusion of the complexity. This means that the findings of a single study (or even a series of studies) is unlikely to be able to accurately represent the complete pathway within the complex system of the natural world. To return to the labyrinth analogy, many scientists conduct research by essentially recreating a model labyrinth in a controlled environment. When the researchers then apply what they learned through the model to the changeable labyrinth in the real world, most likely the model learnings cannot be applied wholesale, but rather provide a small piece of knowledge within a complex puzzle. The extent to which laboratory findings can be replicated in the real world is always a source of uncertainty. This fact can be exploited as evidence that science is not to be trusted by so-called 'science deniers'. The discoveries are not the problem; they are accurate within the context in which they were studied, but when applied to a new context – a more complex system – only parts of the knowledge will transfer.

So why study processes within a context that isn't directly applicable to the real world? Essentially, an artificial environment such as a laboratory serves to suppress or reduce uncertainty, with the aim of creating controlled conditions that allow for studying parts of a complex system. Many researchers use an approach known as 'the scientific method'. This method provides a framework for conducting linear studies that can be reproduced – to help us begin to place the pieces into the three-dimensional, moving puzzle.

The scientific method allows a researcher to explore the model labyrinth in a systematic manner. Scientists are able to test which turn is best, and to begin to reduce the unknowns by following a single path at a time, adding, step by step, to the mapping of the tunnels within the labyrinth. The researcher chooses which path to explore, the method to explore it and how to interpret the results. This scientific method (which in and of itself can be interpreted differently) typically comprises seven sequential steps, starting with an observation and ending with communicating the findings. The steps, and the order in which they are conducted, are the first attempt scientists make to suppress the overwhelming uncertainty present. Similar to the use of checklists in healthcare, this standardised method is an attempt to minimise the uncertainty in complex, changing systems. That's in theory; in reality, every step generates unintended uncertainty that stems from human choice.

Steps 1 and 2: The observation and the question

First, **the researcher makes an observation**, and based on that observation, a research question is developed. The observation and the question is dependent on human choice. What if Dr Fleming, for instance, observed the anomaly of the mould growth and ended up asking how mould had gotten into his petri dishes instead of focusing on the antibacterial properties? The research question and methods may have taken an entirely different path, maybe towards developing better sterilising procedures.

Step 3: Background research

Next, the researcher undertakes background research to better understand what is already known about the topic of interest. This could be viewed as another attempt to minimise uncertainty. Reviewing already published work reduces the breadth of uncertainty related to the research question. The *known knowns* of the field found in the existing literature help the researcher distil a narrower view of the *known unknowns*. Here, again, **the researcher decides** what prior work is relevant, and what conclusions to draw from it.

In the modern scientific community, uncertainty creeps in here because the sheer number of studies produced on a given topic is not something the human brain can contend with easily. Due to this, a researcher's background reading will likely have its own gaps (missing relevant literature) and biases (focusing on what they think is relevant, as opposed to reading all studies with potential relevance to the topic). AI algorithms that search scientific databases, such as elicit.org, can be used to help narrow this down – but humans still need to make choices about what defines the search query, which of the AI-identified papers are relevant to the current study and what the collective studies reveal about the field.

I studied malaria for my PhD. My supervisor was trained as a 'classical' cell biologist, and I was training in cell biology – hence we were both considered outsiders in the field of parasitology. Many parasitologists at the time considered this organism and its cellular mechanisms as unique. This meant some researchers in the field didn't look outside the parasitology literature to explore what was known before asking the research questions or designing the experiments. Enter my supervisor and myself, using classical cell biology, who viewed the parasite as just another organism. Low and behold, (years later) other studies came to support our view – that the parasite had some unique features, but they were variations of classical cell biology, not an entirely novel system. When I first presented my findings to the parasitology community, I was met with scepticism, even some anger, and confusion. The focus of the background research by many in the field meant that they had overlooked some important prior work – the expertise of those doing the research introduced uncertainty in the form of bias into the research process.

Step 4: The hypothesis

From the combination of their observation and their background reading, **the researcher develops a hypothesis**. For Dr Fleming's experiment, the hypothesis may have been: 'If the mould is producing a substance that can kill bacteria, then applying the mould to actively growing bacteria will kill these bacteria.' An effective hypothesis combines the researcher's observations with the knowns of the field.

A human decides what the hypothesis should be, influenced by their personal observations, experiences and background research, as well as their overall research question. Despite being considered an 'objective' approach to research, the scientific method involves human-generated sources of uncertainty.

Step 5: The scientific experiment

The hypothesis is tested by conducting the scientific experiment. Here, to help reduce uncertainty, the researcher typically engages some form of control. For Dr Fleming's experiment, this might have included one petri dish labelled a negative control, which receives nothing at all, or receives the media that the mould survives in – but not the mould. Another dish would contain the mould (this is the experimental setup) and the third dish would be a positive control. The positive control is typically treated with something with a known result (maybe 70 per cent ethanol, as a known killer of bacteria, could be added to one of the petri dishes). Together, controls help test the hypothesis and reduce the uncertainty as to whether the experimental setup is the cause of the bacteria death, or whether some extraneous factor not previously considered could be behind the phenomenon.

Maybe there are multiple experimental setups to explore how different amounts of mould, exposure times and conditions (such as temperature, humidity, movement and so on) all influence bacterial growth in the presence of this mould. Furthermore, each positive result is replicated a number of times to further reduce the chances that the experimental results are due to randomness or serendipity.

The replicability of experiments is a hallmark of the scientific method. I still remember the absolute joy I felt during my postdoctoral fellowship when I saw that two key proteins had bound together in the experimental setup. I ran from lab member to lab member, sharing my excitement and pointing to the result. Within forty-eight hours, the joy had turned to frustration – because my attempt to replicate these results had failed repeatedly. I couldn't reproduce the first observed result, suggesting that it was likely a fluke.

Ultimately, this step is highly focused on uncertainty reduction and suppression strategies in the design – but the human uncertainties can still sneak through.

Steps 6 and 7: Data analysis and reporting

Once the experiment is complete, the data must be analysed and interpreted. The way the analysis is conducted (which statistical method is used) and the meaning of the findings are **affected both by the researcher's knowledge and their experience**. What is considered relevant, and the impact the result has on the field, has subjectivity associated with it.

The final step in the scientific method focuses on sharing the findings and the implications of the research through publications or at conferences. This is meant to help address uncertainties, as the researchers gain feedback and allow others to explore reproducibility in an effort to interrogate the quality of the finding. This step can – no surprises here – introduce further uncertainty. As illustrated by the scientific community's initial dismissal of penicillin, the dogma of the day can change the way the results are prioritised, valued and progressed.

While the scientific method is meant to reduce uncertainty, we can see that science itself – because it is the study of the unknown and is conducted by humans – will always carry with it some uncertainty. But this doesn't mean we can't 'know' things. It just means that scientific proof isn't as simple as many of us (including myself at times during my research career) wish it could be.

The value of scientific 'proof'

Reproducibility and controls help reduce scientific uncertainty. Scientific knowledge is furthered over time when different laboratories explore a subject of interest and add to the literature, resulting in a critical mass of data, where multiple sources build evidence towards scientific proof.

Even with the inclusion of controls, and reproducibility of results in repeat studies, scientists are aware that there may be contexts or variables in which their findings may not apply. This is why researchers use uncertainty code words such as 'suggests', 'seems', 'likely', 'evidence of a correlation between' and so on – particularly when a finding is novel or related to a complex aspect of the field. It doesn't indicate a lack of confidence; it means we know enough that we can identify both the known knowns and the known unknowns – scientists are being transparent about the degrees of certainty and the degrees of uncertainty. Take the example of penicillin. Contemporary medicine is struggling with the overuse of antibiotics, which has led to bacterial resistance to many treatments – including penicillin. This doesn't mean that we can't trust the original experiments that illustrated the efficacy of penicillin; rather, it shows the contextual nature of science. As the world scientists study shifts and changes with time, so too does the applicability of the research findings. This is why scientists can be reluctant to state findings as static facts – they are acknowledging the limitations of the human capacity to study and evaluate an ever-changing system.

You may notice that I use uncertainty code words often in this book to describe my own, and others', research into uncertainty tolerance. I highlight the complexity and nuance in the concept of uncertainty tolerance. Infinite factors interact with our perceptions and experiences in the face of uncertainty stimuli to determine our level of uncertainty tolerance. My use of language, which may seem like equivocation, is acknowledging the beauty of this complexity. There will be contexts where the research on uncertainty tolerance won't apply wholesale – but that doesn't mean the idea of uncertainty tolerance is bogus, just that there are degrees of relevance. I can't know everything about uncertainty tolerance as it relates to every single reader, but I can provide a framework about what is known on the subject.

The uncertainty intrinsic in research was captured well by Nobel laureate Richard Feynman, who I quoted in the chapter opening, in his 1995 National Academy of Sciences address. His quote also inadvertently demonstrates the ubiquity of the archetype of the scientist as male:

> The scientist has a lot of experience with ignorance and doubt and uncertainty, and this experience is of very great importance, I think. When a scientist doesn't know the answer to a problem, he is ignorant. When he has a hunch as to what the result is, he is uncertain. And when he is pretty damn sure of what the result is going to be, he is still in some doubt. Scientific knowledge is a body of statements of varying degrees of certainty – some most unsure, some nearly sure, but none absolutely certain.

The imperfect hierarchy of scientific evidence

Many researchers value the 'evidence hierarchy' as a means to minimise uncertainty to the greatest possible extent. The scientific community has determined that the best evidence of scientific proof is obtained by using methods which (arguably) engage the most 'uncertainty suppression' strategies.

The evidence hierarchy pyramid is divided into two halves. The lower half is devoted to primary studies, which directly observe and study a given topic. The upper half of this pyramid focuses on indirect evidence, which is built by combining these primary studies together, over different contexts and times, to build a broader view of the evidence. This upper portion captures the scientific journey towards truth. These review studies help identify where consensus has been reached in the scientific community and what questions still remain. Essentially, they can give a high-level, broad overview of the *known knowns* and the *unknown unknowns of a given field or topic.*

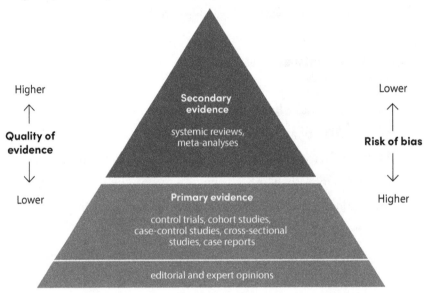

The hierarchy of evidence pyramid: systematic reviews and meta-analyses provide the highest quality of evidence. These reviews are built on evaluating a collection of primary studies. These primary studies can include randomised control trials, cohort studies, case–control studies, cross-sectional studies and case reports. Each type of primary study occupies a rank lower on the quality of evidence scale. Furthermore, each primary study is also proposed to represent an increased risk of research bias and a decrease in supporting evidence. At the bottom of the pyramid are editorials and expert opinions.

Within primary studies, the coveted highest level of research evidence is achieved by research methods known as randomised control trials (RCTs). RCTs are designed to help determine whether an intervention, such as a drug, is efficacious. Because of this, RCTs can be sometimes considered high-risk studies, as humans are often the subjects in these studies when they are related to human health. RCTs can also be used to study educational interventions, but many have heard of these as they relate to drugs or health interventions.

RCTs contain multiple strategies to suppress, minimise and mitigate uncertainty. One feature of RCTs are extensive and thoughtful controls. The negative control subjects are typically administered a placebo, or a pill with similar ingredients to the drug being studied but which doesn't contain the active ingredient. Participant eligibility and selection criteria are used to further 'control' potential uncertainties. People with similar height, diets, weight or hormonal make-up are selected. Yet this tactic has, consciously or unconsciously, led to a historical bias in RCTs, as white males were typically viewed as a study population with the least amount of 'uncertainty'. Some suggest that this is a direct result of the researchers' bias – they themselves tended to be both white and male. The reason most commonly cited, though, is that the complex hormonal changes of cis-gendered females were considered to complicate interpretations of the drug's efficacy. How could scientists determine the impact of the drug if hormones were influencing its metabolism and effects? Instead of viewing this population as important to consider in drug trials, the 'complexity' was viewed as a risk. When we consider that RCTs are the greatest resource of evidence-based medicine in current healthcare practice, we can see that this is more than an academic issue.

What is the theme that keeps emerging? Attempts to suppress uncertainty (especially in an effort to reduce perceptions of risk) simply generate new and different uncertainties – which may sometimes be considered worse than the original uncertainty. In the case of RCTs, uncertainties can be introduced because of the lack of knowledge about a drug on any of the population not included in the study. In other words, attempts to reduce bias in the study protocol end up creating new, but different, biases.

This bias doesn't just extend to human trials; it is present in the research lab and in scientific collections as well. There are multiple reports that male lab animal subjects are chosen over females – often to reduce the uncertainty related to the complexities of XX hormonal changes. This means that before a drug is even tested on humans, its effects are more likely to be studied in only a portion of the model organisms extrapolated to the population who will potentially take it. Thankfully, this is finally beginning to change through shifts in funding rules. It is gratifying that a number of national funding agencies, including the National Institutes of Health in the United States and the National Health and Medical Research Council in Australia, prioritise grant applications for projects that specify animals from both sexes will be studied. A recent article in *Proceedings of the Royal Society B* highlighted that even our understanding of non-human biodiversity is likely biased due to attempts to suppress uncertainty. Scientific collections of specimens from the natural world predominantly contain mammals and birds, as opposed to invertebrates and plants. Furthermore, there are more male specimens than female in most museum collections.

From these examples, we see that uncertainty is introduced into RCTs and other primary research studies when the human researcher decides who or which group to study, how to do so and which uncertainties are suppressed. Ironically, it is often the well-meaning attempts to reduce uncertainty that contribute to the inequality in healthcare: from racial to gender inequities (and the intersection of these) and more.

Search for 'bias in RCTs' in Google Scholar and you will yield over two million papers in every single discipline you could imagine. Again and again, media and research publications underscore that studies tend to exclude cis-gendered females, children, certain races or people of diverse socioeconomic status. In such selection processes, uncertainty isn't eliminated (or arguably even reduced), it is simply redirected.

Children are also placed (unintentionally) in jeopardy when these uncertainty-suppression strategies are used. For example, the side effects of antidepressants were predominantly studied in adults. Some of the results seen in children were reported as 'unexpected'. In the younger population, there was actually an increased risk of suicide with certain antidepressant treatment protocols.

Interestingly, this adverse effect was not identified through a RCT, but rather through the collective observations of researchers and practitioners over time (controls weren't part of this equation). These observations would be considered less important in the evidence hierarchy but were critical to identifying this problem, and led to further research on antidepressants that involved the adolescent population. Other examples include the impacts of lung-cancer treatment on certain research-excluded populations, Alzheimer's disease progression and prevalence research, and the misdiagnosis of heart attack signs and symptoms in cis-gendered women (we saw this one when we met Sally, in Chapter 2). While woman are often cited as the group missing from such studies, there are many more whose absence isn't even noted, such as people who are transgendered or neurodivergent.

Ultimately, individual studies can only ever provide a small piece of the scientific puzzle. A more accurate reporting of primary research studies would clarify the impact of a drug or medication on the entire population. For instance, a change to a more nuanced presentation of study findings – one that is transparent about what is and isn't known – could make a huge difference. Instead of headlines and reports that 'the main heart attack symptoms are chest pain that can radiate to the jaw and down the left arm', it could be communicated that 'heart attack symptoms are chest pain that can radiate to the jaw and down the left arm in cis-gendered men, but other populations haven't yet been studied' so that the public knows in which contexts the study findings apply, and in which they don't.

Surely the top portion of the evidence hierarchy pyramid, which focuses on reviews of primary studies, would more effectively reduce uncertainties than the lower portion? Representing the peak of high-quality evidence are systematic reviews and meta-analyses, which serve as a way to combine broad and disparate information collected from published primary studies in order to obtain a bird's-eye view of the field or discipline in question.

Scientific 'discoveries', we now see, require a series of primary studies done over time and by multiple researchers in diverse settings to build an evidentiary chain towards scientific proof. Systematic reviews and meta-analyses can help to reduce the uncertainty seen in individual studies, as they focus on collating and evaluating diverse sources of evidence. However, uncertainty is still introduced (as with primary studies) through the human choices about what is researched and how it is explored. To reduce this source of uncertainty, the scientific authorities have devised checklist strategies such as the PRISMA (Preferred Reporting Items for Systematic Reviews and Meta-Analyses) guidelines. These are regularly updated. For instance, the 2020 iteration of the

PRISMA guidelines includes upwards of 27 items in a checklist for researchers to adhere to when undertaking reviews.

Those writing PRISMA guidelines are aware that uncertainty still exists despite these best efforts, and so they employ a more flexible approach to standardisations through a series of PRISMA guideline 'extensions'. While systematic reviews and meta-analyses were once the only acceptable forms of review, the inability of these reviews to capture certain diverse data sources or explore some areas (particularly those at the threshold of a new field, where uncertainty is greatest) resulted in an expansion of the types of reviews to include: scoping reviews, realist reviews, semi-systematic reviews, and integrative reviews, among others.

In 2019, a research team composed of PhD candidate Alex Kale and two experts in relationships between uncertainty, reasoning and systems, assistant professor Matthew Kay and associate professor Jessica Hullman, explored where uncertainty appears in research reviews. Their findings underscored how these PRISMA guidelines may help provide boundaries around the uncertainty in the research process, but cannot entirely eliminate it. Their study revealed that scientists are aware of persistent uncertainty: by interviewing a variety of scientists, the authors were able to identify where perceptions of uncertainty occur during the review process and how researchers work to manage it.

Overall, this study found that the majority of researchers attempted to reduce uncertainty by applying the scientific community's accepted rules for conducting research, including using PRISMA guidelines and checklists. If the uncertainty was not reduced using these strategies (for example, if a step in a checklist was ambiguous), researchers tended to acknowledge this uncertainty by identifying possible ways this unknown will impact on their work. Some would acknowledge this uncertainty in the 'limitations' section of their paper. But some researchers completely ignored the uncertainty, often by rationalising it. My team's findings showed a similar pattern: scientists appear to embrace uncertainty to a point, but fall short of entirely rejoicing in the uncertainties despite an awareness that they exist in the research. There appears to be a tension between the excitement of learning about the unexpected and the belief that too much uncertainty will lead to reduced data quality, potentially limiting the meaning and impact of the conclusions.

This 2019 study dug deeper into the sources of this scientific uncertainty and the moderating factors influencing a researcher's tolerance of uncertainty throughout the research process. While uncertainty reduction and mitigation strategies were used across the research phases, early in the research process – when the scientist is developing the research questions and exploring what is known about the field – is when scientists are most likely to acknowledge uncertainty. Here, when all paths are open, uncertainty may be embraced. Researchers are excited about seeing the gap in knowledge as a 'mystery' they get to solve. An acknowledgement of uncertainty appears again when analysing the data, as well as in communicating the results to others. At these stages, scientists seem to seek out the sources of uncertainty in their research so they can develop methods that 'bound' or minimise it. Many also strive not to oversell their results. They see uncertainty at these phases as a way to help them identify their next puzzle, or future directions of the research projects. But while researchers acknowledge certain sources of uncertainty in the analysis stage, this is also when they may attempt to suppress other sources of uncertainty. There is that same tension between wanting to make the uncertainty visible and then work around it to drill down to what their research identified as knowns.

The authors found six domains of moderators, comprising over 20 different factors, that influenced research choices when researchers were faced with uncertainty. The largest domain was 'external constraints', which included resource availability (time, knowledge, equipment, funding and so on). Other domains included code words for uncertainty suppression such as 'consistency', 'reproducibility' and 'standard practice'. Overall, the tension between accepting and suppressing uncertainty appears to be (as in healthcare) predicated on perceptions of risk: some researchers see uncertainty and risk as synonymous and attempt to control both through standardisation.

Dr Joshua W. Clegg, a psychology professor at City University of New York, published a paper aptly titled 'Uncertainty as a fundamental scientific value'. It supports this idea that risk has probably led researchers on a wild goose chase for (arguably impossible) objectivity, resulting in the scientific community becoming hyper-focused on strategies to suppress uncertainty. He posits that, as science became increasingly institutionalised, scientists transitioned from a passionate curiosity in solving life's mysteries towards stability, uniformity and projects that focus on scientific 'certainty'. The finding that 'external constraints' were negatively impacting on researchers' uncertainty tolerance suggests that Clegg's thesis may be accurate.

Science is now big business. Whether or not a researcher is funded determines the job security of the research team. Competitive grant funding impacts on the reputation of a university, the promotion and upward mobility of researchers, and often the perceived value of the findings to the field. Whether or not a researcher is published, and in what journal, drives similar outcomes. Few researchers are supported to publish the failures, with most recognised journals only supporting the 'breakthroughs' and 'discoveries', so it is no wonder that some researchers try to suppress uncertainty despite acknowledging that uncertainty is fundamental to science. Too much honesty about uncertainty in a grant application is likely to reduce the chances of receiving funding. This risk of negative outcome in itself, over time, influences a researcher's approach to managing scientific uncertainty.

Once science becomes a source of risk, uncertainty becomes a threat to be suppressed, just as we saw with healthcare and with business and economics. But suppressing uncertainty in science research can generate bias and misrepresent the fundamentals of the endeavour – it is predicated upon, built from and dependent on uncertainties and unknowns. It also leaves science research open to attacks from those who are anti-science. How can the scientific community address these shortcomings? Could waving a flag of uncertainty help us to relish the beauty and mystery of science?

How to communicate scientific uncertainty (and why we should believe the experts)

One potential solution could be improved communication of scientific uncertainties. A researcher from University of South Florida, Dr Benjamin Djulbegovic, is a physician who specialises in evidence-based medicine. In a 2007 paper, he proposed four different terms to better communicate uncertainty in RCT findings. The terms he recommends, which are drawn partly from the worlds of business and economics, relate to the spectrum of knowns and unknowns naturally present in research – and can serve as a starting point for being more open to and transparent about scientific uncertainty.

At the threshold of a dark labyrinth, when the uncertainty is widespread and the number of unknown unknowns is greatest, Djulbegovic suggests scientists use the term **true ambiguity**. At the start of the pandemic, we had a lot of true ambiguity.

Now, a few years in (and with great speed), we have mapped more of the Covid-19 labyrinth, but we are still unsure about the long-term health implications of the viral infection. Djulbegovic suggests we refer to this stage as **a range of futures**.

When there are clearly defined unknowns, discrete and distinct, that we can test for (sort of like the weather forecaster identifying potential pathways of the cyclone), the term **alternate futures** leaves room for the uncertainty that might occur if what is studied is placed in a different context or complex environment. For example, there is general acceptance of the mode of transmission of Covid-19 (airborne), and we are increasingly aware of the impacts of Covid-19 on the body, but we do not know how these manifest in each population or individual. Long Covid is one area in which we still know very little, though this is changing.

The term **clear-enough future** is proposed to describe types of treatment with clear and significant positive effects on patients. The results of these treatments are so dramatic (Djulbegovic includes penicillin as one instance) that the uncertainty about their impacts are minimal. The mRNA Covid-19 vaccines are a good example here. The results of the trials were very clear, and these vaccines seem efficacious across the broader population with few exceptions.

While science is predicated on uncertainty, this doesn't mean that scientific findings don't hold value and there isn't knowledge gained from research. Quite the opposite, in fact. The key to all of this is the degree to which uncertainty is present in the topic being studied. Those with deep knowledge of the related field of study are the best placed to know which parts of the topic are entirely unknown and which are more settled.

Without expertise, the wrong questions or conclusions could be drawn, or a complex system could be oversimplified to the point of error. Pseudo-science, and pseudo-scientists, can be recognisable because of their tendency to focus on single-study findings, engaging with data that hasn't been reviewed by other experts, and a desire to eliminate any mention of uncertainty. They often present findings as 'sure bets' and 'easy fixes' – they fail to use those uncertainty code words. While this might ease the discomfort that accompanies uncertainty, they can peddle dangerous and false information, such as Covid-19 treatments involving bleach and ivermectin – or the disproven link between vaccines and autism. All of us, as members of the public, need to be on guard for these pseudo-scientific experts.

Scientists need to also improve their communication of uncertainty. If they are more upfront, open and transparent about the permanency of uncertainty in science, society may be better prepared for twists and turns as researchers move between the unknown and the known through the labyrinth of uncertainty.

Many argue that society 'can't handle' or 'won't accept' un-certainty in science, but a recent study in the *Proceedings of the National Academy of Sciences of the United States of America* suggests that communicating uncertainty, contrary to much spec-ulation of those in the field, doesn't necessarily negatively impact people's trust in the science, and may actually help illustrate to the public the stage of uncertainty the field is in.

An important first step in improving communication of uncertainty is a transformation in the way the scientific community itself views uncertainty. Instead of seeing uncertainty as a deficiency, the scientific community needs to be loud and proud about uncertainty being the soil in which scientific discovery grows. Addressing the systemic challenges can start with reframing the way we integrate uncertainty into the most-valued scientific outputs of the scientific community (publications, grant funding and presentations), as well as how we (and the media) translate uncertainty to the general public.

What we see from Dr Djulbegovic's suggested nomenclature changes is that words do matter. They impact the way we perceive what is being communicated, and cast information in a positive or negative light. Dr Djulbegovic, however, does not mention how to address the uncertainty brought in by human researchers themselves – and this may be one of the biggest challenges with uncertainty transparency in science.

In scientific publications (that is, manuscripts), scientists are encouraged to 'hide' this human uncertainty (that is, bias) by using the third person instead of the first. Instead of writing, 'My prior work in X caused me to explore Y', scientists write, 'This research was undertaken to explore Y', concealing the importance of human choice in scientific discovery. By burying the uncertainties related to the researchers' decisions, the scientific community convey findings as more objective than they are – adding to the ammunition that adherents of pseudo-science use against the scientific community. If we switched to writing in first person, we could be more transparent about the researchers' experience and expertise that led them to explore the topic in the way they did – conveying to the public the rigour and knowledge behind the study.

The structure of scientific manuscripts also tends to bury, instead of highlight, uncertainty. Many contemporary journals have word limits. The data is clear that journals prioritise the known knowns (the findings) over the known unknowns, driving most researchers to focus their limited word count on what they found, not on the challenges or uncertainties related to that finding.

Contemporary scientific papers commonly require authors to include a subheading titled 'limitations'. Here, researchers list all the possible known unknowns (and sometimes unknown unknowns). Within this section, they describe the 'flaws' and 'shortcomings' of the work, which can be related to limitations of resources, methodological weakness and so on. These code words for sources of uncertainty are used when the results are likely to vary, or to point to a context in which these results and conclusions may change. The term 'limitations' frames these uncertainties in a negative light, as deficits instead of features. What if, instead, the scientific community embraced these uncertainties as key tenets of research and reframed this section? Instead of limitations, the term could be 'contingencies'. This contingency section could focus on ways the results could be interpreted differently as new data, different methods or other uncertainties become known. As the complex environment changes – as the labyrinth moves in time and space – the authors are setting up different ways to interpret their results and areas of future work.

Scientists could further normalise this positive perspective on uncertainty by dedicating time in their presentations or media discussions to covering contingencies. Here the researcher (and hopefully, in time, the media) can make the 'what if' and known unknowns a regular element of the coverage. By doing this the scientific community can better prepare listeners for the inevitable changeability of science at the cusp of bold ideas.

Another way that the scientific community can embrace uncertainty is to draw upon diverse bodies of knowledge and on experts outside of those identified through a traditional colonial lens (the archetypal scientist). Victor Steffensen describes himself as an 'Indigenous filmmaker, musician, and consultant reapplying traditional knowledge into the changing world and today's society'. His fascinating book *Fire Country* is a terrific example of how traditional knowledges can help solve contemporary scientific challenges. His book focuses on Indigenous land management techniques, including cultural burning practices, to help manage Australian bushfires. Many other First Nations writers, including Karlie Noon, Bruce Pascoe and Corey Tutt, explore how Aboriginal people pioneered agriculture, engineering, astronomy and architecture on the Australian continent long before the colonials claimed this science.

Even opening up the language of science could help. Currently, the international standard language for communicating science is English. A 2019 *Nature* article discusses the challenges with the adoption of a single global language. Not only does this make it more challenging for those who learned English as a second (or third, or fourth) language to be accepted in the scientific community and become perceived as an expert, but it makes it more challenging to convey scientific findings to a global audience, in a world with many diverse languages and ways of engaging. Furthermore, a focus on 'certainty' in scientific English creates yet another barrier for competent and capable researchers.

By being open to diverse sources of scientific discovery (and diversity in communication strategies) we benefit the field of research as well as society. While progress is being made in this area, peer-reviewed research from multiple fields suggest that privilege is still given to researchers in the Northern Hemisphere and those of developed nations. Imagine how much further ahead we would be in tackling tangled problems (problems that appear near impossible to solve because of the extent of uncertainty) if we didn't rely on a single method of research; if we embraced the uncertainty of diverse knowledge and research methods; if we challenged the quest narrative; and if we regularly discussed scientific findings in languages other than English?

In this uncertainty-inclusive scientific community, funding applications would be benchmarked for authors' ability to highlight (not hide from) potential sources of uncertainty. Grants would be awarded not just on the potential to 'close a gap' in knowledge, but also on the researcher's capacity to open up new pathways to guide the whole field towards the next exploration, the next big question. That is, funding would be based not only on what you hope to do but also on how you will generate the next source of uncertainty for the field to explore. Funding bodies with these principles at their heart would support diverse ways of understanding the world, not just traditional colonial methods. The funding would reward creativity, diversity and lateral thinking – all intricately linked to uncertainty.

Tied to each of these recommendations is the overall recognition of scientists. If we continue to rely on metrics that prioritise 'certainty' (citations, publications, grant funding and awards), scientists will be driven to suppress uncertainty or risk their careers. Alternatively, if the scientific community instead values those who ask big questions, who challenge the knowns by seeking out the unknowns, we may be able to collectively re-orient towards a form of scientific discovery that tolerates uncertainty.

A 2022 *Nature* editorial underscores concerns that 'research-assessment systems are too narrow in what they measure', favouring those that 'secure large grants, publish in high-impact journals, or register patents, at the expense of high-quality research that does not meet these criteria'. But which measures could help in this re-orientation towards uncertainty tolerance? The European University Association and Science Europe – made up of 350 organisations – took four years to develop solutions to this challenge. Their answers lie with valuing qualitative metrics of research, such as 'the standard of leadership and mentorship' and public availability of science through 'data sharing and collaboration' – essentially, focusing on recognising more than just the publication, but the impact of this work on the scientific and general community.

Experts aren't just old white men in lab coats

Critically, the way we reward science needs to change. By reinforcing the idea that individuals are responsible for profound scientific discoveries, the scientific community is suggesting that only certain expert opinions are relevant, and that uncertainty, missteps and failures aren't part of the scientific process.

This concern is a very real experience for many researchers. A 2022 paper led by Dr Eunji Kim, an assistant professor of political science, discusses the impacts of being a researcher from a background that is traditionally underrepresented in academia. It opens with a powerful statement:

The coauthors, all scholars of color, have found that in addition to the general difficulties encountered in field research, our identities in particular pose other challenges. Our expertise, objectivity, and status are doubted, occasionally met by muted enthusiasm from research participants. When researcher identity defies the expectations of a typical profile of an academic affiliated with North American- or European-based institutions (i.e., particularly white and male), it has important implications for inferences drawn from field experiments.

The text goes on to illustrate the judgements made about the authors' credibility by those they must work with in order to undertake the research. This questioning of their credibility, despite the fact that they are just as knowledgeable as their archetypal-scientist counterparts, leads to multiple challenges throughout the research process. These range from participant recruitment challenges, stemming from institutions being less likely to engage with these researchers; to data collection challenges, due to changes in behaviour when participants see a researcher that doesn't 'fit the expectation'; to inappropriate behaviour towards the researchers, including incessant questioning on aspects irrelevant to the study, such as about the researchers' ancestry. Finally, because these researchers don't represent the archetypal scientist, they face perceptions that they are somehow biased.

If we change the narrative, exposing the reality that a diverse set of researchers have worked jointly to accomplish some of the greatest scientific discoveries, we may be able to (over time) adjust societies' perceptions of who is credible. By sidelining those such as Rosalind Franklin and Alice Augusta Ball for so long, we have created a scenario where only a select group of scientists are perceived as experts. The stories of scientific discovery will be no less rich by sharing more of the full story of discovery. After all, if the discovery of penicillin really ended with Dr Fleming, we wouldn't have had a viable pill.

There are multiple studies illustrating that science development and discovery is best undertaken by diverse teams, both big and small. A small research team published a paper in *Nature* in 2019. They explored manuscripts published between 1900 and 2014 and compared the data to patents and software developments related to this research to examine the relationship between team size and make-up, and the impact on scientific progress and disruption. They found that smaller teams tended to have the greatest impact on innovation, while larger teams seem to help with the incremental development of these disruptive ideas. A pair of researchers also published a paper exploring the role of teams in research impact back in 2013, and found that research by diverse teams had the greatest impact on the field. Adding to this, gender diversity also appears to lead to enhanced research outcomes. In fact, inclusivity and diversity in science seems almost imperative in a world filled with growing uncertainty.

If the scientific community can strive to convey to the public more about the known and the unknowns, recognise the collaborative nature of discovery, and be more transparent about the human choice in research, we can reassure the public that uncertainty is a foundation of building knowledge, not something to fear or challenge. We can better convey the joy and wonder that is science research – in all its complexity.

As individuals and as a society, we need to look to a person's credentials and their experience, not the colour of their skin or their gender, when determining who is an expert.

At the same time, to avoid being sucked in by pseudo-scientists, we must always weight up the sources of scientific knowledge experts espouse. Have they published peer-reviewed work in this area? Are they a graduate from a reputable institution? Are they recognised as an expert by their peers? Are they highlighting the uncertainty? If the answer is no to many of these questions, or if the so-called expert is selling certainty, we should be wary of what they peddle. It may be pseudo-science.

The media also plays a role in this scientific conversation on uncertainty. There aren't always two sides to every story. The media needs to use the same criteria that the scientific community and the public do to identify an expert. Those without expertise in the field, who aren't seen as experts by many of their peers, should not be given airtime. Experts also have a responsibility to convey the uncertainty and resist the idea of reducing scientific findings to simplistic statements of fact. A recent paper suggested that **less than 25 per cent** of media coverage of science research communicated the uncertainty related to the findings. When the media did discuss uncertainty, there was a focus on linking 'uncertainty' and 'risk to health', suggesting again that these are synonymous, painting uncertainty as a negative aspect of scientific research. The media has a crucial role in communicating scientific research to the public. A simple change in the narrative could go a long way towards helping the public embrace (even seek out) the uncertainty intrinsic in science.

How to teach future scientists to maintain wonder

To change society's view of science, we also need to change the way we teach science. We have generated an educational system in the sciences that reinforces quest narratives, linear pathways to discovery and limited diversity in what is considered 'evidence'. One of the research participants in my team's studies highlighted how we need to replicate the uncertainty present in the scientific process through the way we teach science to students:

> In biology, we're always exposed to natural variation ... whenever we do biological experiments, there's never going to be a certainty ... the hypothesis is there, but you're never ever going to get a perfect result ... that's why we have to do a lot of technical replicates. I think one way I prepare students for that is I introduce them to natural variation, and variation in biological phenomena.

How we teach science is a key factor in whether the uncertainty inspires wonder or fear. Sissy Wong, associate professor of curriculum and instruction at the University of Houston, writes that 'we have to change how we teach science for the future'. While her focus is on preparing the future energy workforce, her words align with what my team has found in our research on general science education. Two paragraphs are worth quoting:

> ... traditional science teaching involves having students complete scientific experiments by following the scientific method, which is a prescribed and linear process. This usually includes asking a question, doing background research, constructing a hypothesis, testing the hypothesis with an experiment, analysing data, drawing a conclusion and reporting results. But scientists don't actually follow a linear process when they are working, and teaching science in this way gives the impression that science is stagnant and prescriptive.
>
> Another example is the notion that science is objective, with no role for personal beliefs and creativity. This is inaccurate.

Scientists around the world hold many different points of view that influence how they pursue their work. It is not realistic to believe that backgrounds do not influence how we see the world. That's especially important to remember as the nation's student body becomes more diverse.

These are the challenges that traditional ways of teaching science generate in preparing the future workforce for the innovation and curiosity that is required of them. Furthermore, this type of teaching reinforces that science seeks certainty, and that truth is discovered through linear and stepwise progression. Instead, teaching science using the methods outlined in Chapter 1, fostering a tolerance of uncertainty among students, could be a better way forward.

What might this look like? What if, for instance, we introduced a panel of researchers into the classroom, each member exploring the topic that the students are learning about in science? The 'experts' will likely debate their findings and use diverse methods to draw conclusions – which conveys the changeability of science in both the findings and the interpretation of these findings. Or, instead of having all students use the scientific method to explore a research question, teachers could introduce diverse methods for exploring the world around us, including interpretivist methods such as qualitative methods and design thinking approaches (these are described in Chapter 3). Both of these methods serve to help get at truth by centring the perspectives of those experiencing it and embracing the researcher's role in the research process. Instead of narrowing in on a single hypothesis, these studies focus on research questions to better see what may be currently unseen from the single perspective of the researcher.

Students would begin to appreciate the complexity of research methods, and also learn how different approaches can all help lead to knowledge and truth.

How to embrace and learn from scientific uncertainty

For those of us who went through a traditional schooling system, where the linear scientific method was king and where heroes made discoveries: don't give up. There are steps that we can all take to learn to embrace the positives that uncertainty brings to scientific research.

First, it is important to remember that uncertainty in research doesn't equate to a lack of truth or valuable knowledge. In order to discover something new, what is being researched, by its very nature, needs to be uncertain. *Sense about Science: Making Sense of Uncertainty*, a pamphlet about uncertainty in science written by a diverse team of dozens of researchers, states: 'Uncertainty is normal currency in scientific research. Scientific research works on the basis that there are things we don't know.' To manage this, scientists build experiments to learn more, and over time the scientific community accumulates an evidentiary chain leading to what they term 'settled science'. Just like most legal cases are built on a series of evidence, so too are the 'truths' of science.

Instead of distrusting science because it changes, we can look to this as important evidence that the scientific process is working, and that we are learning something we didn't know before. Be wary of people who are sure about something that is still debated by experts. Researchers with relevant expertise will be able to identify the sources of uncertainty and the extent of this uncertainty in their research – so ask them to do this. Ask them what they think is still not known about the topic, or what new information would change their minds about their conclusion. Experts will be able to describe the history that led to the current finding and name the 'contingencies'. Trust in science should occur because of uncertainty, not despite it.

Importantly, think beyond the quest narrative. Value experts even if they look different from the archetypal science hero. There are many credible voices that don't match the profiles of the voices the fables have taught us to listen to. And they are telling us some of the best information.

To quote Joshua W. Clegg again: 'Uncertainty and ambiguity are not scientific problems to be solved – they are the very genesis of knowledge.' By reorienting the narratives of scientific discovery away from certainty and heroic figures towards the messiness, the collaborations, the repetitions filled with knowns and unknowns, we can better understand scientific truths and why they matter.

5

CULTURAL FLUENCY

How uncertainty tolerance can foster
diversity and social inclusion

*'My humanity is bound up in yours, for we can only be
human together.'* —DESMOND TUTU

n June 2022, pop artist Lizzo dropped a new and much-anticipated song: 'Grrrls'. Yet many listeners noticed something off about it, and soon took to social media to comment. Sets of 280 characters of anger, frustration, indignation, emojis and gifs were soon burning up the internet.

What got so many so worked up? A single word with multiple meanings. A powerful word. 'Grrrls' contained the word 'spaz' in the opening stanza. If you are a British or an Australian reader of this book, you might have just read this word with the same viscerally negative reaction that those social media posters had. If, however, you hail from the United States, you might be wondering what the fuss is all about. This is because the term has developed a culturally diverse meaning.

The term 'spaz' is deeply offensive in many cultures, but not in all. Because of the global release, this song sparked a controversy that soon made headlines. It followed a similar outcry in 2006, when Tiger Woods used the term to describe a poor execution of a putt during the US Masters tournament. And it didn't stop there: in August 2022, only two months after Lizzo's release of 'Grrrls', Beyoncé dropped 'Heated', which also used the term and sparked anew disapproval over the use of this 'ableist' language.

American linguist and language commentator Ben Zimmer explores when and how the term diverged so significantly in meaning between developed nations in a blog post. While in the United Kingdom and Australia this term is a shortened version of the word 'spastic', historically used to describe a condition that alters the nervous system's motor control, in the United States society a related but different term, 'spaz out', was used to mean acting eccentrically or 'losing your cool'.

Zimmer cites a BBC report that traces how, in the United Kingdom, the terms 'spaz' and 'spazmo' experienced a resurgence in connection with Joey Deacon in the early 1980s. Deacon, who had cerebral palsy, became a household name when he appeared on a children's show, *Blue Peter*. Although the show presented Deacon as an inspiration, the outcome did not match the intention. BBC journalist Damon Rose notes:

> Never was [the term's] potency or currency so big as when the programme featured Joey Deacon in the early 1980s, believing the story of a 60-year-old man with cerebral palsy overcoming the odds would touch the hearts of under-12s.
>
> Oh, how wrong. It unleashed a monster. [Several negative slurs for disabled people] became familiar phrases that year and were still being used years later by gurning children in the playground. 'Spaz' became synonymous with useless incompetence – the type you see in disabled people [sic] portrayed

badly on TV ... Not long after, The Spastics Society famously changed its name to Scope.

In Australia, for some time the term was considered, along with other now derogatory words such as 'cretin', as the appropriate way to describe someone with a disability. In recognition of the negative connotation of the word, the Australian organisation historically named The Spastic Centre, originally formed by parents of children with cerebral palsy, renamed itself Cerebral Palsy Alliance in 2011, with an ABC report highlighting the 'increasing sensitivities to the word spastic' as a key driver in the name change.

The connotation in the United States, however, is very different. Zimmer explains that by 1956 the term had come to represent someone who follows the rules or is a 'square'. This use of the American term 'spaz' was epitomised in pop culture by Chas the Spaz, a character on *Saturday Night Live* in the late 1970s. Picture Steve Martin, pants hiked up, with a partially tucked-in button-down shirt and a sweater vest, a pocket protector and glasses – this was the archetypal 'spaz' made popular by the iconic sketch comedy show.

Fast-forward to 2022. People from the United Kingdom and Australia called out Lizzo's song for being ableist and offensive, while people in the United States were confused, with many only learning that the term was offensive *because* of the controversy. Lizzo's response on Twitter and Instagram was swift:

It's been brought to my attention that there is a harmful word in my new song 'GRRLS'. Let me make one thing clear: I never want to promote derogatory language. As a fat black woman in America, I've had many hurtful words used against me so I [under]stand the power words can have (whether intentionally or in my case, unintentionally). I'm proud to say there's a new version of GRRLS with a lyric change. This is the result of me listening and taking action. As an influential artist I'm dedicated to being part of the change I've been waiting to see in the world. Xoxo, Lizzo.

Lizzo's reply included elements of uncertainty tolerance: she left space for others' views, acknowledging that her experience with the term could be different from theirs; she listened, learned and acted. And her story reveals something important: just because there is increased exposure to other ways of living, being and knowing through social media and news outlets doesn't mean that we can always *know* what it is like to live and work in another location, or how to fit in with another culture's values. When we leave space for uncertainty, we leave space for understanding others' cultures, as opposed to condemning them; we leave space for exploration, as opposed to stagnating; we leave space for curiosity, as opposed to being know-it-alls. Whether or not we perceive uncertainty when we encounter cultures or societies outside our own, and the way we respond, is informed by our uncertainty tolerance.

The concept of uncertainty tolerance (or ambiguity tolerance) is most often credited to a Polish-Austrian psychologist, Else Frenkel-Brunswik, who identified it in 1949. Frenkel-Brunswik's affinity for the subject may have related to her own experiences with cultural challenges and differences. A Polish Jew, Frenkel-Brunswik moved with her family to Vienna to escape anti-Semitism in 1914. Here she completed a PhD in psychology.

Forced to migrate to the United States due to anti-Semitism in 1938, she joined the University of California at Berkeley, focusing her research on personality and psychoanalysis. Frenkel-Brunswik soon after began studying anti-Semitism, from which the concept of uncertainty tolerance was born. Her work, later supported by others, revealed that individuals' *in*tolerance of uncertainty was linked to racial prejudice. Those with less tolerance for ambiguity were more inclined to perceive differences as threats and conflicts, leading to feelings of anxiety and discomfort – manifesting as racism. So the concept of uncertainty tolerance itself is, at least in part, rooted in cultural differences and variabilities – and our responses to them.

One of the key factors that makes us human is our ability to navigate uncertainty. Our choices, autonomy, conflicts, variabilities and diversity can all be sources of uncertainty. Opinions and values different from our own can stimulate our perceptions of uncertainty, and how we respond to these uncertainties is influenced by our level of uncertainty tolerance.

Our uncertainty tolerance, as we see throughout this book, is also influenced by moderating factors. In a societal context this can include our socioeconomic position, our geographical location, our employment status, our position within the social hierarchy, our own cultural and ethnic influences, and so on. Despite one of the very factors that make us human is that we all have to navigate uncertainty, many of us seem to struggle with, and battle against, it.

When uncertainty appears in community settings and challenges us in ways that we perceive as placing our social position or livelihoods at risk, we can tend towards actions that seek to suppress these varied sources of uncertainty. Fear of uncertainty is a natural response to the unknown, but our uncertainty tolerance may determine how strongly, and in what way, we act in the face of social uncertainties.

In a globalised world, individuals can interact with, observe or learn about cultures different from their own with greater ease. With each observation and encounter with an unfamiliar or unknown community, an individual's level of uncertainty tolerance is challenged. Will we view these differences with fear or curiosity, and how can we learn to become more tolerant and less afraid?

Cultural flavours

I have had my own small experience of cultural discomfort and linguistics. Despite moving from the United States to another English-speaking country, I struggled with communication during my first year in Australia, especially when words were the same as those I used at home but had an entirely different mean- ing. Several times I put in an order for 'two to three' egg rolls per person for a workplace lunch. The caterer always questioned our administrative assistant on this order: 'Are you sure? That is **a lot** of egg rolls.' To which my colleague would reply: 'Yes, I am sure. I triple-checked with Michelle.' The caterer would shrug. 'Well, we do get lots of positive reviews about our egg rolls, I guess.'

Finally, after the third time this happened, I followed up with my colleague. Why weren't we receiving the egg rolls I ordered? She looked confused, and pointed to the large pile of curried- egg salad wraps. We realised what had happened, and I began to laugh. I was ordering two to three wraps per person *every single meeting*. I intended to order two to three spring rolls per person, but in the United States, spring rolls are the Australian equivalent of rice paper rolls, and egg rolls are spring rolls. So, as in Lizzo's case, the same term had an entirely different meaning between cultures.

I clearly had a lot to learn, and language was just the start.

When I was contemplating moving, a friend who had also lived and worked abroad said to me, 'You will never feel Australian, but once you have lived there, you will also never be "American" again.' These words continue to ring true for me, even now, as an Australian citizen. My understanding of the Aussie way of life continues to improve, to the point that I actually miss the daily experience of learning something new. But I am still conscious that I don't quite understand the nuances that come with being raised in the culture I live in, and I have accepted that I may never know them all.

Cultures are networks of complex nuances developed over time through an iterative process of human interactions. I attribute my own curiosity to learn about Australian culture in helping me, at least a little bit, understand the values and priorities of my newfound home. A tolerance of the cultural uncertainties is what allowed me to bridge the gap between my own experiences and knowledge, and those that shape the realities of others.

It stands to reason that if we develop uncertainty tolerance at an individual level, we can create a society with an appetite for learning what we don't currently know, and for being curious about something (or someone) who may be different from what we are familiar with.

But uncertainty tolerance is only one piece of a complex puzzle in managing cultural uncertainties. Award-winning American sociologist Dr Milton Bennett's developmental model for 'intercultural sensitivity' addresses this relationship between cultural uncertainty and the spectrum of potential responses more deeply.

Intercultural sensitivity – it's all relative

In 2022, a Twitter user shared a map of different countries and whether a foreigner 'will receive food as a guest at someone's house'. Scandinavian countries (Finland, Sweden and Norway) were categorised as 'very unlikely to give you food'. In Australia, showing up empty-handed or not offering something to a guest might seem wrong. The responses to the map illustrated that many interpreted the rightness and wrongness of offering food to guests based on their own cultural norms, while others viewed the differences with curiosity, seeking to better understand the cultural motivations that lead to these differences. These two approaches represent the spectrum of intercultural sensitivity.

The spectrum of intercultural sensitivity explains the varied responses that individuals can have when we encounter unfamiliar cultures. At one end of this spectrum, there are those who respond to cultural differences from a perspective of **ethnocentrism**, and at the other end is **ethonorelativism**.

Ethnocentrism takes meaning from two Ancient Greek words: 'ethnos', translating roughly to nation/people/group, and 'kentron', meaning centre. It refers to the assumption that one's culture and experiences are at the centre of the world. Ethnocentrism, as a sociological concept, explains the tendencies of some people to use their own cultural experiences, knowledge and views as the frame of reference in understanding others' cultures. Ethnocentrism within Dr Bennett's model is typically represented by people who encounter a cultural practice different from their own and respond with denial, defence or minimisation – denial being at the extreme end of the spectrum and minimisation a more tempered response.

Some of the social media responses even *after* Lizzo addressed the controversy are good examples of ethnocentrism. Despite sharing evidence that there were indeed cultural differences in the meaning of the term 'spaz', some readers from countries where the term was viewed as derogatory and ableist still thought that Lizzo 'should have known' the negative interpretations – suggesting that they were only seeing through their own cultural lens.

Ethnocentrism has been present throughout history. In a 2017 article by journalist Natalie Cromb, ethnocentrism features in a description of how James Cook's view of Aboriginal peoples was shaped by the British explorer's own culture. By viewing Aboriginal cultural practices in reference to his own colonial social standards, Cook's views of what represented 'civility' informed his conclusion that Aboriginal (and Torres Strait Islander) peoples' cultural practices were 'savage'.

But ethnocentrism isn't just a historical cultural lens. In Australia, our present-day understanding of many social norms can also be ethnocentric, due to our colonial history. Take for instance the case of Moale James, a Brisbane woman who was refused entry to the bar Hey Chica! because of her cultural tattoos; the venue had a blanket policy against allowing patrons with 'head and face tattoos' because such inkings are often linked to those in 'criminal organisations'. Because tattoos were viewed through an Anglo-Celtic lens, a policy was made that took away the relevant context of another culture. Moale, who has Papua New Guinean heritage, told the ABC, 'I just want a little bit of empathy, a little bit more respect', qualities central to the other end of the intercultural sensitivity spectrum: ethnorelativism.

Ethnorelativism is when, instead of weighing others' cultures against one's own, we enter uncertain territory with respect, built on an acknowledgement of the differences between the world we are familiar with and that with which we are not. Instead of viewing another's culture through the lens of our own, ethnorelativism assumes that the frame of reference for evaluating another's culture is variable and context-dependent. The responses to cultural differences in this part of Bennet's spectrum include acceptance, adaptation and integration.

Lizzo's response to the backlash over her song epitomises each of these stages. She doesn't assume that the connotations of the word 'spaz' in her own culture represent a universal truth. She adapts by learning a different understanding of the term and integrating this new understanding into her lyrics and way of thinking, culminating in the release of a modified song and a public acknowledgement of her unintentional insensitivity.

A strong disposition towards ethnocentrism is linked to racism, stereotypes and xenophobia, because such individuals tend to value their own cultural experiences as 'right', and others' as 'wrong' or 'lesser'. Where ethnorelativism seeks to avoid judgement of another's culture, ethnocentrism tends to support a hierarchical social structure with in-groups and out-groups, where in-groups tend to be defined by the dominant cultural group in a given society. There is a tendency to view others' cultural differences as something to fear, or feel threatened by: key characteristics of uncertainty intolerance. In contrast, ethnorelativism seeks to acknowledge the uncertainty present in understanding different cultures and embraces it. Cultural unknowns are, in this frame of reference, not something to fear, supress, deny or minimise, but rather to be curious about, to seek to understand. In order to engage in ethnorelativism, one needs to be appropriately tolerant of uncertainty.

While the spectrum of intercultural sensitivity is a sociological tool designed to map individual responses, we can see similar patterns at the societal level, too. Systems that are set up to value and support certain individuals at the expense of valuing and supporting those that don't fit the societal norm (as an extension of ethnocentrism) may generate social injustices and inequities. There are theories, such as cultural hegemony and uncertainty-identity theory, that suggest our collective intolerance of uncertainty is what allows those in power to maintain and reinforce their dominance. When we don't question the norms, when we contribute to the suppression of those deemed 'atypical', we are individually displaying features of uncertainty intolerance and collectively contributing to a societal structure that is also intolerant of uncertainty.

A body of evidence is beginning to develop that shows building one's uncertainty tolerance counteracts ethnocentrism. Take the 2013 study of approximately 300 undergraduate students enrolled in an intercultural communication course. They found that participants' uncertainty tolerance, and not cultural knowledge, was the driver for reducing ethnocentrism. They suggest that a way to address ethnocentrism is developing a lifelong curriculum that fosters uncertainty tolerance.

This doesn't mean that ethnorelativism, and the uncertainty it provokes, is necessarily comfortable. As we will see in more detail in the final chapter, discomfort nearly always accompanies perceptions of uncertainty – particularly in high-risk contexts. What I am suggesting is that the discomfort we experience in seeking to understand other cultures can be a sign that we are in a position to learn and grow. It helps us focus energy away from assumptions and towards empathy. In contrast, if we do not challenge ourselves to be uncomfortable in the uncertainty – if we instead seek comfort by using our own experiences (certainties) as the benchmark – we do not allow ourselves to see the visible and invisible social structures that contribute to bias and social inequities. In addition, we risk creating new ones.

Diverging from the norms

Our challenge as individuals within a social system is that our attempts to suppress uncertainty can introduce bias and reduce equity for others. We saw this in the chapter on healthcare, where people who do not display the typical symptoms of a medical condition can be mistreated or misdiagnosed, sometimes with fatal consequences. We also saw this in the chapter on science, where randomised controlled trials often include only one group but make conclusions about the broader and more diverse populations. When we consider the cumulative effects of these attempts to suppress uncertainty across our society, we understand how the quest for certainty can inadvertently introduce and reinforce favouritism and exclusion in our social system, contributing to social injustices across the population.

Let's look at some examples. Two children, Olivia and Sandra, both seem to generate uncertainty in those around them. They have trouble fitting in and often feel like outsiders. Here is how their stories play out.

Olivia is outgoing and curious. She loves to be outside in the bush, exploring and learning. She asks lots of questions, and her family obliges by indulging her never-ceasing inquisitiveness and unquenchable excitement for learning.

Fast-forward, and Olivia is now in Grade 3. Her 'adventurous' spirit is now a source of trouble. She has difficulty in following the classroom rules, reading the directions in class exercises and staying quiet in reading time, which the rest of the class seems to do with ease. Her report cards come home with comments like 'smart, but can't sit still', 'intelligent but disorganised' and 'lots of potential but talks too much'. She is beginning to feel like she is different from others, and she starts to struggle in school. By Grade 7 she is feeling anxious about school, and particularly struggles with tests and exams. She knows the content, but she has trouble focusing under pressure – she consistently misreads the questions, despite knowing the answers. Her marks don't reflect her knowledge. The teachers don't seem to know how to manage her – she's smart but atypical, and they treat her classroom behaviour as a problem to be solved.

Fast-forward again, to high school. Olivia isn't sure what her options are for the future. Her marks on exams are still sub-par – try as she might, she just seems unable to focus. Whenever she notices a student turn over their test to finish early, she becomes stressed and panicky – why can't she figure out what is being asked? She is keenly aware of her shortcomings; they have been drilled into her by her teachers through the feedback, constant hushing, grades and even movies and books that portray how a 'normal' girl is supposed to think and act.

At the last possible minute, Olivia applies to university. She chooses a degree in biological sciences, because it was the only subject she did well in consistently, and it seems like this degree could give her a career that might allow her to engage her natural curiosity and her passion to understand the world around her.

At university, she fails her first exam. Olivia's chemistry professor requires all students who fail an exam to set up a meeting to discuss study strategies. She contacts the professor to set up the meeting. She waits nervously outside his office.

The meeting, despite her anxiety beforehand, is the first time she doesn't feel out of place. The professor acknowledges that she knows the content well when called on in class; in fact, he is impressed by her depth of understanding. Olivia explains that this exam result isn't unique – nearly every exam mark doesn't seem to reflect her actual knowledge.

She cries a little, and he is kind about it. He suggests that she meet with the university disability services. She's hesitant. This is just who she is, right? It is who she has always been. She isn't disabled, just bad at exams, right? But she knows that something needs to change – so she follows her professor's advice and makes the appointment.

The disability services team suggests testing for a learning disability. She is shocked at this suggestion but eventually acquiesces. When the results are revealed by the psychologist, she sums them up in a single sentence: 'You are the poster child for ADHD.' These seven words challenge Olivia's entire identity. She isn't sure which of her thoughts and behaviours are truly 'hers' and which are symptoms of ADHD.

The diagnosis has some benefits. Olivia gains access to support, including a study mentor and extra time on exams – allowing her the opportunity to take her time and read the questions properly. With these provisions, she excels at university, but the earlier grades and feedback from her teachers continue to haunt her. She looks to apply for PhD positions, and one professor, for whom she works as a teaching assistant, tells her bluntly, 'You will never make it.' He seems to only see the disability, not her abilities.

She reaches out to a mentor, who says, 'You'll never get in if you don't apply.' She applies, but is wracked with anxiety about whether she has the ability and whether she will fit in if she does get offered a place. After a lifetime of being told *she* is the problem, it is hard to shake these negative thoughts.

To her shock (less so to those who know and love her), she gets into good schools, goes on to a post-doc, and embarks on a career in academia. Her creative, inquisitive brain – features of ADHD – make her an ideal fit within the academy. The change-ability and diversity of roles and the focus on exploration are all things Olivia enjoys, and this fits the norm expected in her chosen field. She teaches, as well as undertaking research, and wins awards in recognition of her teaching. She uses many of the lessons she learned about being outside the norm at school to support her students.

With ADHD there often comes an ability to hyper-focus on certain activities of interest – for Olivia, this is writing. She relishes solving thorny problems and maintains her passion for learning. This combination of attributes, despite being frowned upon when she was growing up, are the very ones which see her rise through the ranks in her career.

In the workplace, she strives to mirror others' behaviours, but the toll can leave her exhausted. She has learned how to organise and plan in a way that works for her neurodivergent brain and that allows her to perform to the neurotypical standards of the work-force she is in – indeed, many even comment on her excellence in organisation and follow-through.

But her inability to conform to the status quo (she often still speaks out of turn or asks too many challenging ques-tions – qualities that are often frowned upon as 'unprofessional') means that she is sometimes overlooked for positions despite hav-ing the capacity, intellect and passion to excel in these roles. She is also permanently plagued with the anxiety that has resulted from decades of negative feedback. The uncertainty she generates, just by being herself, results in societal responses that serve to suppress the uncertainty – and Olivia internalises this as negative self-talk.

This is ultimately a good-news story, but let's stop and wonder. What if the world around primary-school Olivia wasn't so uncertain about where she fit in, and was focused on exploring how to help her earlier on? What if her teachers were curious about how her non-typical thinking and behaviour could add value to the classroom, instead of suppressing it? Would she feel more confident now? Would she have been more successful? Would other careers have been open to her?

Let's now consider someone whose skin tone means they can't hide their 'difference' as Olivia tried to. Sandra is African-American. She is brought up in a nice town in the middle of the United States. She attends the local private school. She notices that most of her peers are white but doesn't (at this stage) think much about it.

As an all-girls institution, the school seems very progressive, with a significant part of the curriculum focused on women's rights. When she is young, everyone in her class seems to get along and she doesn't notice any of the potentially negative interactions with her peers that her parents have prepared her for.

Fast-forward to Grade 6. Sandra, now eleven, can choose where to sit in the cafeteria. She starts to notice that the room is now broken into peer groups whose members look like each other. Instead of sitting with her year-level class, she is now seated with children of all different grades who look like her.

One day a student, a white girl, shows up with the 'N' word written on her arm. Sandra and her friends tell this girl with the racial slur on her arm how offensive this is. The girl quickly dismisses these concerns, saying, 'It's just a joke. I don't mean anything by it.' Sandra and her friends report the incident to the principal – it means something to them. Sandra is sure there will be consequences for this girl; after all, the school is very progressive. This school surely doesn't tolerate inequities and injustice. Sandra knows that if she showed up with a racial slur on her arm, she would be suspended.

The school's response? They gather Sandra and her group of friends, a counsellor and the offending student. The counsellor asks Sandra and her group of friends to explain to the student how the situation made them feel – leaving the victims to educate the perpetrator. The offending student apologies. From the school's point of view, the matter is now resolved.

This is Sandra's first true lesson that things are different for people with her skin tone. She is in disbelief. Sandra thought that society was far too advanced to have such things happen. She begins to see the invisible hierarchy in society, and her role towards the bottom. She starts to question her safety at the school, and within society, now realising that feminism may not be inclusive of rights for *all* women.

Years later, and it is time for Sandra to apply for college. The school is interested in tracking students' applications as a promotional tool for future students and their parents. Now Sandra is starting to feel the pressure of being a 'model minority'. The school pays for her to apply to all of the Ivy League universities, but she also wants to apply to one of the historically Black colleges. She applies to the Ivy Leagues, alongside Spelman, a women's liberal arts college in Atlanta, Georgia. Sandra gets into several schools and narrows her decision down to Spelman and an Ivy League school. She gets a full scholarship to the Ivy League school, and likes its program better, but she worries that going there, rather than Spelman, will make her less of a role model for other African-American women. Sandra is left feeling anxious about her decision to attend the Ivy League school.

She is successful at university and applies to medical school. Applications require a letter of support from a single medical-school adviser at Sandra's university. She sets up a meeting with this adviser, looking forward to sitting down with him and planning her next steps.

She enters his room, expecting him to go through her transcripts. But the meeting didn't go as planned. The only thing she recalls about this meeting now is the adviser saying, 'You are not doctor material.' She begins to question everything about her dreams, her aspirations.

She doesn't apply for medical school, and instead goes to work in a research laboratory. After two years in this role, her research mentor says, 'This is ridiculous – you need to apply to medical school. This is your future – this is your passion.' Sandra takes the advice and applies to an MD/PhD dual-degree pathway, and gets in. She is, however, now two years behind her peers in training.

During her dual higher degree, Sandra is subjected to many comments suggesting that her grades weren't what got her there, but her minority status. She also learns that other Black medical students who had the same supervisor were all told they weren't 'doctor material', reinforcing what Sandra suspected all along – that the colour of her skin was defining her role in society.

She makes it through her dual degree on intellect and sheer persistence, but there are scars. The anxiety grows, alongside a negative internal dialogue. Her mind seems to play a refrain on repeat: 'I don't belong here.'

Sandra begins her specialist training in the field of reproductive endocrinology. While a trainee doctor, she meets a white patient who exclaims, 'Oh, they're letting Negros be doctors now. We've come so far as a society!' She is frequently mistaken for a nurse. Once, despite Sandra introducing herself to a patient and their family as the doctor, ordering the tests, discussing the results and planning the next steps, when a white male nurse entered the room, the grandmother turned to him and said, 'Your nurse, Sandra, is doing such a great job.'

Sandra completes her training years and transitions to a clinical academic position. Instead of being offered the position of assistant professor, with a large remuneration package, as every single other trainee (all white) accepted into this department was given before her, she is told, 'We think you need more time to develop, so we want to give you an extra year before your tenure clock starts.' This translates to a role that carries the same responsibilities as the higher-ranked position but at lower pay. The only advantage is the tenure clock hasn't started ticking. She accepts, but again at a price.

The internal dialogue of 'I don't belong here' begins to change to 'Is this where I want to be?' Sandra can't seem to figure out if she just doesn't fit in, or if the system in which she works within is making her feel out of place. She contemplates leaving academic medicine, but with the ever-present pressure to be a role model, she feels forced to consider more than herself in this equation. If she leaves the profession, she could make it more difficult for up-and-coming non-white doctors. The burden of a decision for her wellbeing versus what she feels she owes those around her seems insurmountable, and she is struggling to see where her choices end and society's pressures begin.

She is now years behind her colleagues, financially worse off and questioning whether she should stay in her career. I ask her how she isn't just angry all the time. She responds: 'Getting angry is a form of privilege. If you are in the minority group and you get angry, guns blazing, you are immediately dismissed and the problem doesn't get solved.'

Both case studies, which are real stories, illustrate how invisible barriers permeate society. Olivia and Sandra are both aware at a young age that their place in society creates uncertainty, and that they will have to work harder than others. And because they don't seem to fit particular expectations, sometimes those around them aren't sure how to interact with them. Our society deals with divergence and differences by developing averages and policing norms, or by expelling those who generate uncertainty to the periphery, to the margins.

As a society, we need to form a balance between rejoicing in individual difference and developing shared frames of reference for what is or isn't acceptable within a given group or culture. Without societal boundaries and structures, we risk anarchy. But how do we define difference in this context? What about the girl who grows up wanting to use her brother's microscope but is told 'no' and handed a doll instead? Or the boy who wants to wear dresses? Or the transgender child who wants to compete in the Olympics one day?

Our culture's visible and invisible barriers and standards serve us certainty about how to act, think and feel (for example, girls play with dolls and boys wear pants). If you don't fit the norm, if you generate uncertainty in those around you, the message is that there is something wrong with *you* and the expectation is that you will work to fit in, to assimilate.

Those determining the social norms have a vested interest in continuing to support structures that result in marginalisation and disadvantage. But it's not all on the powerful. So how do we know when to question a cultural norm, and when to embrace it?

Power imbalance – cultural hegemony and the status quo

When society supports a single dominant culture, our reference point for what is 'certain' and 'uncertain' comes from that group in power. We can end up seeing that which doesn't conform as 'wrong' instead of just 'different', and diverse behaviours and experiences (uncertainties) as needing to be suppressed instead of embraced. The concept of cultural hegemony can explain how these norms are established and reinforced throughout a given society.

Antonio Gramsci, a Marxist intellectual and one of 'the most influential theorists of the twentieth century', is credited with developing the theory of cultural hegemony. The term 'hegemony' refers to the influences of the dominant group in a given culture or society. Unlike power obtained through control and coercion (picture military dictatorships), it refers to a more insidious, almost invisible power exerted over others. Cultural hegemony is when those in power manipulate the culture of a society so that their worldview becomes the accepted cultural norm.

The status quo is maintained through social institutions. The media, churches, schools and government bodies all send overt and subliminal messages about what is acceptable and unacceptable in our society and define the status quo. While we think we have autonomy, cultural hegemony theory suggests that our decisions about what is and isn't acceptable is actually a 'Hobson's choice' – only one option is viable, so there isn't really a choice at all. While those in power may have greater autonomy, for the rest, the message is often 'conform or you will be pushed out'.

For instance, if Sandra chooses to get angry about the inequities she has experienced in the medical system, she is likely to be further marginalised – considered 'difficult', even unprofessional, and perhaps overlooked for positions. She either conforms to social norms or she pays the price – a Hobson's choice.

An unlikely representation of the theory of cultural hegemony is the 1999 sci-fi film *The Matrix*. The rebel leader, Morpheus, gives the protagonist, Neo, a choice between two pills, representing a choice between the comfort of ordinary experience or the uncertainty of learning a life-changing truth: 'You take the blue pill ... the story ends, you wake up in your bed and believe whatever you want to believe. You take the red pill ... you stay in Wonderland, and I show you how deep the rabbit hole goes.' Essentially, submitting to the cultural hegemony means choosing the blue pill. In this context, we all go about our lives, making choices and decisions and believing we have control – but we are influenced by the 'the matrix', or a set of hidden rules made by those in power.

Cultural hegemony isn't just in the movies. In 2022, a viral TikTok video showed how cultural expectations can influence our perceptions of certain individuals. A narrative often reinforced in western societies is that older women who are alone are also lonely. We are led to believe that if a woman is alone after fifty, something in her life must have gone wrong: she was too focused on her job, she is a bereaved widow, she is a divorcee yearning for a partner. And god forbid she sits alone at a shopping centre café and has a cup of coffee! This cultural myth gained new currency after a TikTok influencer recorded himself asking a woman 'sitting alone in a food court' to hold some flowers. He walks away, leaving her looking down at the surprise gift in amusement. 'I hope this made her day better,' the video is titled. Many of the 39.9 thousand comments it attracted were patronising in the extreme: 'That was the most beautiful act of kindness … you made my day'; 'She needed that man'; 'I think her husband died so that's why she cried and she thinks that her husband gave her the flowers'. These commentators didn't stop to ask if this woman *was* lonely or unhappy – or if she was simply enjoying some alone time. Their (conscious or unconscious) bias about lonely old women was an example of buying into cultural hegemony.

In fact, Maree, the woman in the video, told ABC Radio that she was just enjoying a coffee. She didn't want the flowers and tried to return them. Then she was saddled with figuring out how to take them home on the tram, along with her shopping. 'It's a patronising assumption that women, especially older women, will be thrilled by some random stranger giving them flowers,' she said. In an ageist, patriarchal culture, older woman who are alone are deemed to be weak, vulnerable, incompetent, or all of the above.

Social media is just one medium that can perpetuate the status quo, reinforcing the values and priorities of the dominant culture. We are strongly influenced by the norms represented in pop culture, advertisements, films, and so on. Gramsci argues that each of these influencers serve to reinforce cultural hegemony, affecting the choices we make. Cultural hegemony nudges us away from curiosity about difference and towards accepting (without question) the status quo. Think about the last time you filled out a form. Women often first select our marital status: Ms, Mrs or Miss. All men are Mr. This common practice is reinforcing a subtext that women's role in society is based on our relationship to others. Me? I always choose Dr – just to avoid the expected tick-boxes.

In 2006, James Forrest and Kevin Dunn wrote an interesting article about the forms of cultural hegemony in the United

States, Canada, New Zealand and Australia. They found that each of these colonised societies represented Anglican privilege – the dominant group in these societies tended to promote the social norms and rules aligned with the Anglican Church. Professor Rowan Strong, an expert on the history of the Anglican Church and an ordained priest, agrees. In a 2003 paper outlining the role of institutional Anglicanism in Australian society, he suggests that the Church of England was the dominant cultural influence up until World War II, shaping Australia towards more conservative values, and argues that only recently is this orientation being challenged to include Aboriginal and Torres Strait Islander peoples' ways of knowing, being and doing. Australia's cultural hegemony – our views on what defines dress, language, manners, work ethic and social behaviours – is strongly aligned with the philosophy of the Church of England. Different societies will have different forms of cultural hegemony based on their histories and politics, even if some share a clear orientation in a particular direction.

Dunn and Forrest suggest that Anglican cultural hegemony has persisted more strongly in Australia than in other colonised societies, partly because of the White Australia policy – a set of discriminatory policies that disadvantaged First Nations Australians and non-white immigrants, and continued until the 1970s. They suggest that political initiatives around multiculturalism introduced by prime minister Paul Keating in the 1990s were a significant challenge to the existing hegemony and generated nationwide uncertainty about Australia's social structure. Whether or not they are right to be so laudatory of Keating, it is true that a renewed focus on multiculturalism kicked off a public conversation: what are the social norms in a multicultural society?

How do we interact with those unfamiliar to us? We began to collectively ask who we were and what we stood for.

When faced with cultural uncertainty, every individual has a choice. We can be curious and inquisitive, or we can be fearful and look for ways to generate cultural 'certainty'. Forrest and Dunn propose that the election of social conservative John Howard as Australia's 25th prime minister and the rise of the One Nation party were responses to this challenge to our social identity and the existing cultural hegemony. When faced with uncertainty, brought on by particular socially progressive policies, some Australians responded with intolerance.

In contemporary Australian society, as in most other western societies, progress to question and dismantle the predominant cultural hegemony is slow. We still see evidence of the systemic architecture of the patriarchal British colonialism that has shaped Australian culture for some time. This evidence includes: the gap in life expectancy between Indigenous and non-Indigenous Australians, inequities in healthcare across different populations, incarceration demographics, and so on. Why is cultural hegemony so persistent, even as society (arguably) becomes more tolerant? The answer may lie in our individual tolerance for uncertainty.

How a lack of uncertainty tolerance can reinforce social inequities

In a 2013 article, Drs Michael Hogg and Janice Adelman discuss uncertainty–identity theory. Feelings of belonging and identity within a social group are fundamental to humanity, they argue. When you don't feel that you belong, the consequences are so great that individuals are 'highly motivated to belong to and be accepted' by a community.

Uncertainty–identity theory suggests that when faced with high levels of uncertainty about our identities (self-related uncertainty), we seek to suppress the discomfort by finding a way to fit in. Self-related uncertainty manifests when we are questioning central aspects of our lives, such as who we are and where our place is. For instance, when I first arrived in Australia, I was unsure of where I fit in both personally and professionally – my self-related uncertainty was high. Hogg has written that 'people identify with social groups to decrease feelings of self-related uncertainty'. Our threshold for self-related uncertainty drives our decisions about aligning with, joining and supporting different social groups.

Hogg suggests that self-related uncertainty is usually accompanied by what we anatomists know as an autonomic response – more generally known as the fight-or-flight response. We all have a similar biological reaction when faced with an unknown challenge, whether we perceive it as a threat or an adventure. Imagine that you are going on a blind date. This experience is filled with uncertainty, and you may wonder how you should act on the date. So, in addition to the uncertainty of an unknown situation, you begin to doubt yourself as well. Many of us in this situation will have the same biological responses. We'll feel a surge of adrenaline, represented by a rapid heartbeat, butterflies and maybe changes in skin colour (some blush, others go pale). Some will interpret these physiological responses as indications of excitement, while most will perceive these same reactions as manifestations of anxiety. Hogg proposes that when we experience these biophysical responses along with significant self-related uncertainty, most of us experience enough discomfort (anxiety and fear) that we seek to 'resolve and reduce' this uncertainty quickly. We actively search for ways to minimise the negative feelings.

An example of this could be when you start a new job. You might be unsure about how to dress or what the social norms are at the company. You might question whether you will fit in. You may perceive the risks of not fitting in to be higher if you are on probation or in a test period, where your ability to work within your team is being evaluated, and if your livelihood depends on this job. It would be hard for many of us, in these circumstances, to simply embrace the uncertainty and not seek to assimilate into the workplace culture.

Hogg suggests that when we are questioning our identity in new cultural contexts, we search for certainty about who we are and how we should think, feel and act – and we look outside ourselves to find the answers. In the workplace example, we might be reserved on the first few days (or weeks) of a new job, spending our time observing the dynamics, figuring out the expectations and determining the best approach for success in this environment. By embedding ourselves within a group with clear 'rules', we can generate a more certain definition of who we are and what we should value in this culture. We reduce our self-related uncertainty. Hogg observes that aligning with a social group 'renders the social world and one's place within it relatively predictable and allows one to plan effective action, avoid harm, know who to trust, and ... know how one should feel and behave'.

While seeking belonging is a natural impulse, it can be a dangerous game when we hope to resolve our self-related uncertainty solely by joining a social group. If our capacity to tolerate self-related uncertainty is lower, we may find that we end up seeking out rigid social groups with more inflexible rules and boundaries – extrinsic certainty – to counteract the ineffective management of our intrinsic uncertainty. In doing so, we may generate societal ripples that lead to injustice. When we identify strongly with one group, it means that we end up *not* identifying with another group. Many refer to this as 'othering'.

Humans tend towards generating social structures made of 'us and them' groups. The group we align with becomes 'us', and the outsiders become 'them'. We receive validation from the group when we align with the group dynamics and expected reactions, and this serves to reduce our self-related uncertainty. The more rigid the group structure is, the clearer to those within and outside it what is appropriate and inappropriate. The rules generate a soothing mirage of certainty about our identities: we know when we fit in and when we don't, allaying the autonomic responses we interpret as trepidation about our place in the world.

Social groups with clear boundaries, a predictable social order and a focus on common goals, shared values and priorities tend to alleviate feelings of self-related uncertainty to a greater extent. Such groups are likely to demonstrate, Hogg states, 'internal homogeneity and consensus, orthodox and ideological belief systems and associated worldview, ritualised practices, profound ethnocentrism, hierarchical structure, and emphatic leadership'. The less we can pacify our own self-related uncertainty, the greater the likelihood that we seek reassurance through such groups. Hogg's research has illustrated that 'uncertainty significantly strengthen[s] identification with and behavioural intensions for the radical group' and weakens individuals' association with more tempered, moderate social groups. In times of great uncertainty, many find comfort in social groups who display ethnocentrism and maintain the cultural hegemony – members don't need to guess who is in and who is out.

How do we avoid this? After all, social groups hold a necessary function within society. Feeling connected supports our overall wellbeing, both mentally and physically. It is a key factor in combatting loneliness. When we share our good news with others, our listeners mirror this positivity through non-verbal cues (smiling, raised eyebrows) and biological responses (heart rate and hormone release), generating social connections. Dr Dave Smallen, an expert in human connectedness, states, 'Human beings rely on these positive, synchronous moments as a basic connecting force beginning in infancy, and people continue to seek out synchronous interactions throughout life.' My own team's research into how university students manage uncertainty in the classroom suggests that their learning community (their fellow students and teachers) can help them to navigate the destabilising features associated with uncertainty. In these collective studies, students who navigate uncertainty on their own are more likely to show behaviours that represent an intolerance to uncertainty – they disengage.

The answer is, in part, about increasing our own tolerance of uncertainty, so that we are less reliant on a sense of group belonging to ease our personal anxieties. The way we manage feelings of self-uncertainty appears to influence the type of social groups we align with. If we are more tolerant of uncertainty, we may be more likely to align with groups where the borders of what is acceptable and unacceptable are more fluid. It becomes a virtuous circle: people who align with more moderate social groups are also less likely to depend on the group's rules and dynamics to bolster their sense of self.

In short, our own level of uncertainty tolerance influences our ability to manage both extrinsic and intrinsic sources of uncertainty, which in turn influences the very fabric of society.

DIFFERENT SOCIAL GROUPS

When we feel great self-uncertainty, we can feel unsure of where to fit into society and how to act (panel 1). Those who are more tolerant of uncertainty are able to manage these feelings and engage loosely with social groups (panel 2). Those less tolerant of uncertainty require external sources of certainty and tend to align with more rigid social groups (panel 3).

Cultural literacy and other ways to change the system

Literacy has two standard definitions: 'the ability to read and write' and 'competence or knowledge in a specified area'. Typically, we focus more on the first than on the second, but both are important. Appreciating and translating cultures outside of the one we are raised in, through what is termed **cultural literacy**, is an essential skill in a globalised world. (And let me pause here: while the definition of 'culture' is hotly contested, the definition I use stems from a book titled *Among Us: Essays on Identity, Belonging, and Intercultural Competence*, co-authored by Dr Jolene Koester, whose research focuses on communication studies, and emeritus professor Myron Lustig, an expert in intercultural communication. They state: 'Culture is a learned set of shared interpretations about beliefs, values, norms, and social practices that affects the behaviours of a relatively large group of people.')

Cultural literacy is the ability to read and understand a culture different from our own. To quote a chapter published by my team:

'Broadly speaking, [cultural literacy] refers to how we can become "literate" in difference, and how we can learn to "read" and interpret differences (particularly in terms of culture), in the same way we can learn to read and interpret a text.' The purpose of practising cultural literacy is to help with making meaning, in a broad sense, about cultures we are unfamiliar with and uncertain about. Cultural literacy can be used to build our tolerance of uncertainty; our own social norms and practices determine what we consider to be socially 'uncertain' and influences how we will respond.

This idea of 'reading and interpreting differences' in the same way a text may be read and interpreted can be challenging for some. Dr García Ochoa, a lecturer in the school of languages, literatures, cultures and linguistics at Monash University, recalls being at an interdisciplinary conference and explaining cultural literacy:

I spoke about 'readability' – how we can read and interpret the world much like we do a text, which is an idea grounded in semiotics. One of the attendants stood up, visibly shaken, and said, "An apple is an apple! You eat an apple, you cook an apple, but you can't *read* an apple – that makes no sense!" I tried to explain to him that the word 'apple', and the notion of an apple, could have different meanings in different contexts, and these notions of literacy and readability refer to how accurately we can perceive and interpret those different meanings: when someone says "apple", are they referring to the IT brand? The Big Apple? Buenos Aires? The apple of discord, or one of the golden apples of the Hesperides? Are we talking about Genesis and the fall of [hu]mankind? Red or green? Gala or Granny Smith? Cultural literacy, at its core, is being open to learning how to "read" the symbols you encounter – including an apple – and understand the meaning of these symbols to a given individual, a culture or society as a whole.

García Ochoa and his colleague, Professor Sarah McDonald, highlight how the three phases of cultural literacy (described in more detail in Chapter 1) are accompanied by discomfort and destabilisation. Cultural literacy might be good, but it doesn't necessarily feel good. Again, we encounter the theme of learning to be comfortable with discomfort when we navigate uncertainty.

Cultural literacy could be an agent in the fight against cultural hegemony. García Ochoa and McDonald emphasise that cultural literacy includes an openness to differences and a curiosity to explore one's own bias and preconceptions – and seek to move past them. They suggest that by engaging in cultural literacy we can develop empathy and a greater awareness of those different from us, which in turn supports our capacity to challenge the norms.

For this to be successful, though, workplaces need to be designed to acknowledge the value in challenging practices and norms. Those in power need to foster an environment that encourages others to question apparent certainties, and create a space acknowledging discomfort, supporting individuals as they manage it. Those who question should be supported to initiate change and even empowered to lead this change. Instead of asking employees to conform to workplace norms, leaders need to find ways of acknowledging and celebrating the differences and supporting interrogations of the status quo.

Sound big-picture? It's not. How *do* we get individuals to bring about change in, and across, a social system purpose-built to be resistant to it? Let's remember that society is built on and by individuals. So if we can motivate individuals to identify biases, call out assumptions and build supportive work environments, we can bring about real, systemic change because individuals make up those systems. There is this cyclical loop between individual responsibility to change this system and the responsibility of the system to change.

Researching and writing this chapter provoked anxiety, as it triggered my own uncertainties. I questioned my authority to write on this topic. I may be neurodiverse, but as a white, educated academic, I am also in a position of great privilege. I encountered great discomfort stepping into experiences, theories and knowledges in which I don't consider myself an expert, and for which I often lack personal context. I truly came to understand, though, that the discomfort is the starting point of learning and change. The uncertainty-induced discomfort tells me that I am at the threshold of a new way of thinking. It is a necessary precursor to transformation.

Through writing this chapter, I have come to realise the critically important role that those in positions of power have. Exclusion, bias and inequity is linked to the tension between certainty and uncertainty. Social inequities persist because many individuals in privileged positions are not questioning, probing, interrogating or challenging the visible and invisible power structures woven into our societies, and they aren't taking actions to change these structures. We are the ones that need to critically reflect on the decisions we make and the systems we support, both consciously and subconsciously. As leaders in a workplace, teachers in a classroom or government policymakers, we need to be at the front of the pack, leaning into the uncertainty, questioning norms, developing our tolerance of uncertainty and supporting its development in those around us. We need to be setting the example and helping to create systems that are also uncertainty-tolerant.

We saw how invisible barriers influenced Olivia and Sandra's personal and professional trajectories. Their lives were changed by the actions and decisions of those who unconsciously and consciously bought into social 'certainties'. What if those of us with privilege paused and questioned what norms were leading our decision-making, and why the norms were there in the first place? What if the medical school supervisor, or the university president, had interrogated the supervisor's tendency to label every Black pre-med student as 'not doctor material'? What if Olivia's and Sandra's teachers taught their peers to see them as adding value rather than as insufficient? Would these two very capable individuals have excelled even more, instead of questioning themselves?

Some leaders may ask themselves privately what the motivation is to question the structures around them, if they are benefiting from the privilege afforded them. There are at least two solid reasons: justice and efficiency.

I was privileged to listen to a talk by Dr Amanuel Elias, a research fellow at Deakin University and an expert in race relations, ethnic inequalities and cultural diversity. He is also the co-author of a book, *Racism in Australia Today*. In the talk, Dr Elias made two cases for addressing systemic inequities: the social justice case and the business case. The social justice case is simple: highlighting the inequities and modes of oppression in our culture, and working to address these, is an ethical issue – a moral imperative.

But while we'd all like to think this case is enough to sway us alone, history shows us that it is not. The business case makes an argument based on economics: the estimated financial burden to society resulting from impacts on health, including mental-health disability. Inefficiencies 'make social injustices economically unjustifiable'.

Dr Elias's research, jointly funded by VicHealth and the Australian Human Rights Commission, gives us an insight into the costly 'business' of racism. When the final results of this study were published in 2021 in the *Journal of Bioethical Inquiry*, it revealed some shocking facts and figures. He calculated that 'racial discrimination cost the Australian economy an estimated $37.9 billion, or 3.02 per cent of the GDP, each year in the decade from 2001–11'. To put this in perspective, from March 2021 to March 2022, the Australian GDP rose 3.3 per cent.

What this groundbreaking research illustrates is that in an unjust society, we are all losers. As the Race Discrimination Commissioner at the time, professor of sociology and political theory Tim Soutphommasane, put it: 'We know there is a human cost to racism in our society. Now we know something about the economic cost of it.'

If you are a leader, regardless of which case motivates you to step towards change, do it. When those in positions of power set a culture of critical reflection, they can support others to do the same and they have the power to act to address the inequities when they are found. When individuals take personal responsibility for change, the system itself is changed. Instead of viewing the system as an insurmountable obstacle, leaders can reframe this to say: I have the power to create a culture that is supportive of critical reflection, that is tolerant of uncertainty, and that can work to address the inequities when they are found. We can't put it all on the system, when the system is composed of people like us.

This view could be considered idealistic. Traditionally, those in power have had a hard time sustaining the motivation to change a system that has benefitted them – but if we resist change, that doesn't mean that everything will stay the same.

While social media can be used to perpetuate the status quo – as the 2020 documentary *The Social Dilemma* shows brilliantly – we have all seen how it can also challenge the status quo. When the masses take the controls on communication, sharing uncensored and open thoughts, more people, including those who have been pushed to the margins, can begin to publicly question what many in society take for granted. They can challenge social norms directly and with immediacy – unlike traditional forms of mass media such as network television, newspapers and Hollywood films.

We saw this with both Me Too and Black Lives Matter. Social media can actually change the script on societal attitudes and behaviours, making visible the embedded social structures of patriarchy, racism and classism. Any of us with an internet connection have the power to call out prejudice and support diverse and novel voices with varied experiences. If the cacophony of dissonant voices solidifies into a movement, the mainstream media will report on it, amplifying the impact across society. Social media allows us to harness people power and generate movements at speed and scale. So while those in power can and should lead social change, if we don't, perhaps those without such a platform will build one for themselves.

The influence of these societal campaigns is palpable. Watching a documentary about a 1990s music festival recently, there were several times I had to press pause. The interviewees' recollections included experiences of rampant misogyny, allegations of sexual assault, and signage that objectified women. The norms of the 1990s came flooding back to me as I watched the clips and heard the stories. I shuddered as I recalled that at the time we, as women, were still too often (this was after third-wave feminism, mind you) encouraged to feel flattered by unwanted sexual advances and accept misogyny as an inevitable facet of masculinity. Women who didn't conform were marginalised as 'frigid' and 'man-haters'. I wondered how could we, how could I, have thought this way? I know that I was conditioned to think this way, to adopt the values of those in power, through movies, headlines, policies and regulations in American society. I am almost ashamed to admit that it wasn't until the Me Too movement that I began to interrogate this deeply. The shadow over my vision began to clear. I started to question all sorts of norms around women and confront my own unconscious biases. Not everyone was like me – there were some whose vision was always clear – but social media provided the chance to build a loud, undeniable chorus of voices that couldn't be ignored. It brought down perpetrators and questioned the systems that supported them. While there is still much work to be done in dismantling the patriarchy, social media generated a sense of energy and an appetite for change across a broad swathe of society, including those in government.

Australian lawyer and writer Nyadol Nyuon OAM wrote in a July 2022 article that 'the powerful and those disproportionately privileged by the status quo have a problem with social media', because this unfettered access means that everyone can speak up and be heard – and those in the privileged positions aren't able to censor or control them. Throughout history, literacy was often denied to those who were marginalised and oppressed. Why? To help prevent uprisings against the status quo. Just having platforms to communicate and question norms in our society isn't enough, but it is a good start.

How to embrace diversity and reduce social inequity in the workplace

Workplaces and society both benefit from diversity. It turns out that supporting diversity – religious, political, educational, cultural, socioeconomic, or related to gender, sexual orientation or abilities – comes with concrete, measurable benefits. A 2018 study about diversity in business shows that 'companies with more diverse management teams' have higher revenue from innovation. Even if innovation isn't your company's goal, a 2018 Deloitte Millennial Survey shows that the millennial workforce (soon to be the majority workforce) *expects* and *seeks* a diverse workplace. Sandra now looks at the websites of future employers to see how diverse the team she will be working for is – and this research suggests that she is not alone. Social justice, diversity and inclusion have become critically important considerations for both the wellbeing of the workforce and the economy.

When the motivation to change exists, the challenges workplaces often face is where and how to start enacting this change. One of the biggest lessons I have learned through my research is that there isn't a checklist for managing social uncertainty – as much as we might like one. But here are some examples from the literature, and some potential recommendations for how you and your team could help improve workplace diversity and social equity.

A first step that leaders can take to address inequity is to audit their organisation and explore the level of diversity present. Dr Mandy Truong, a public health researcher who focuses on health equity and racism, offers details about what this audit should include. There needs to be a review of 'an organisation's culture, policies and practices – for example, the complaints mechanism and governance structures'. 'Workplaces need to openly explore who has a seat at the table in making decisions, who only gets the crumbs falling off the table and who is absent entirely. Who are the gatekeepers, and who are the enablers of change? We need both carrots and sticks to incentivise people to take action, and hold people accountable for inaction,' she says. In other words, workplaces need to first take stock of who they are and what skillsets they have (or lack) before moving towards greater equity and inclusion.

Once the organisation is aware of their starting point, the next step is to develop a culture of inclusion. This can start with considering the language we use at work. In recognising the gendered, ableist and classist terms we may use without thinking, we can begin to interrogate the biases in language that reinforce the status quo.

A team from *Harvard Business Review*, led by behavioural science researcher Odessa S. Hamilton, examined how language can have a direct impact on workplace diversity and inclusion when it comes to hiring. Their 2022 article opens with a simple comparison of leadership terms such as 'chair' versus 'chairman' or 'boss' versus 'lady boss'. When you read each term, how does your mental picture change? The use of gendered terms in job ads, including the addition of 'man', 'woman', 'boy' or 'girl' to a neutral term, generates an expectation of both the person in the role and the attributes of that leader. The result? Applicants who don't fit the stereotype of, say, a 'chairman' are less likely to apply. The authors note that 'simply using the word "competitive" has been shown to deter more women than men from applying for a job' and that 'gender-biased language may also contribute to the underrepresentation of women in STEM'. Language that represents historically male figures, such as a 'ninja', has similar effects.

By being more considered in the language we choose, and by selecting words that don't signal the type of person 'expected' in a certain role, we leave room for uncertainty. We provide an opening to encourage new types of candidates. And we resist unconsciously preselecting certain familiar types of candidates.

Hamilton and her team make four recommendations for supporting diversity through language:

1. Reviewing job postings to ensure neutral language

2. Devising a list of forbidden words

3. Creating a guide to inclusive language

4. Engaging ambassadors – highly visible people in the company – to support inclusive language initiatives (leveraging the 'messenger effect').

Let's look at each of these in turn.

The authors highlight the work of another team who explored, in greater depth, how language in job advertisements can indicate a workplace that perpetuates ageism, racism and ableism. This team, led by Dr Ian Burn, an economist at the University of Liverpool whose research focuses on the 'economics of discrimination and its impact on labour market outcomes and health', found that language can signal to potential applicants an 'ideal candidate' with certain physical or intellectual abilities. For instance, coded words like 'dependable', 'experienced' and 'reliable' may suggest that an older individual is sought; younger generations need not apply. Burn and team suggest that replacing these terms with words like 'adaptable' and 'creative' can signal to a wider audience that they should apply. Removing gendered 'he' and 'she' pronouns in favour of 'they' can also broaden a job advertisement's appeal. After all, language is more than a series of letters – it's a product and a symbol of culture, as Dr García Ochoa makes clear when discussing cultural literacy.

I noted an example of this covert signalling in a job ad I came across for a local hospital. The video that accompanied the ad showed a diverse workplace, with staff and patients of different ages, genders and cultural backgrounds. At first glance, this ad looked like it 'ticked all the boxes' of inclusion. But something caught my eye. I noticed that the moment the words 'respected and trusted' appeared on the screen, to describe the values and priorities of the hiring hospital, the image accompanying these words was a middle-aged white man. Cultural hegemony is everywhere.

It is a challenge to be consistently aware of our own unconscious bias, no matter how hard we try. There are tools available to help, which look for exclusionary hidden language. Hamilton and her team recommend the Google open-source browser extension Gender Blinder, which removes implicitly gendered terms from webpages. Part of being tolerant of uncertainty, however, is humility – an acknowledgement that we may not have all the answers and may need to employ those with expertise (including lived experience). One strategy could be engaging a consultant expert in inclusive language or cognitive psychology to help identify language that could be deterring applicants – and (unintentionally) making those in the workplace feel less included. Another idea is asking employees for feedback, maybe asking whether they would apply for the position based on the draft job ad, and what would encourage or discourage them from this decision. This may produce confirmation bias, as existing employees may provide feedback that supports the status quo. An alternative is forming a focus group or a committee of community leaders from outside of your workplace. These participants should be paid and acknowledged appropriately for their time.

Once a position is advertised, take stock of who is applying, as Dr Truong suggests. Many job portals allow applicants to elect to give demographic information, such as gender, geographical location, age, religion and so on. By drawing on this data, an employer can determine if a job ad is attracting a wide range of applicants. If it is not, employers may want to consider ways to reach certain target demographics, such as advertising in different places or promoting the role in a creative way – or reaching out to community leaders.

Supporting diversity in hiring is an important step, but it is one aspect of many in creating an inclusive workplace. I know of situations where people are hired as part of programs aimed at fostering diversity, but once through the door, the system remains unchanged, and they quickly see that only employees who conform to the status quo thrive. Again, language helps with creating ongoing and sustained inclusion in the workplace. A list of 'off-limits language' makes the organisation's values and standards clear to all. This list could include 'do not use' words that carry stigma, or phrases that reinforce stereotypes, as well as, of course, racial slurs and other forms of abuse. Reviewing emails, project documents, style guides and written and verbal feedback is a good way to help identify which words and phrases in use should be designated as 'off-limits'.

Who should lead this review? The leaders of the company, with expert consultants and community partners as needed.

When leaders ensure that inclusive values are communicated and standards held to, it signals to the entire workforce that this is a core business practice. To bring about a sustained culture change, building in opportunities for critical reflection is a wise idea. Most workplaces review presentations, reports and professional documents to ensure they convey ideas accurately. A review for inclusive language could be integrated into this process. Instead of reviewing a presentation for accuracy and then for inclusive language, these could be done at the same time. This reinforces that equity and inclusion aren't add-ons but essential practice.

Given that language and culture are constantly changing, instructive documents, such as the 'off-limits language' list, should be considered a living document, so all employees feel empowered to discuss words and language, and to contribute to this list – ideally on a fairly regular basis (particularly at the start of a culture change). The more regularity to such practices, the more they become the new norm – and over time, cultural literacy and tolerance of uncertainty is woven into the fabric of the organisation.

To support a workplace-wide change, a sort of 'inclusivity handbook' could also be developed, say Hamilton and her team, with lists of helpful resources to support individuals. Such resources could include, for instance, the Australian Government's guide to language inclusivity, focused on 'language that is culturally appropriate and respectful of the diversity of Australia's peoples', including Aboriginal and Torres Strait Islander peoples, those with a disability and non-binary individuals. The handbook could include the organisation's cultural principles, 'before' and 'after' job ads to provide examples of hiring protocols, steps to take when an employee believes they have experienced discrimination, and ways to contribute to improving the inclusivity of the workplace (more on this below). If the handbook is widely distributed and referred to frequently, it becomes a genuine tool for long-term change. Like the 'off-limits language' list, this handbook would be updated regularly.

Finally, Hamilton and her team recommend appointing change champions. These people would help motivate others around inclusivity and diversity initiatives, and also provide gentle feedback on aspects of diversity and inclusion as the organisation develops new policies and procedures. Building a workplace culture that provides positive recognition to those who speak up and contribute to the diversity framework is essential. Rather than a singular 'diversity and inclusion award', for instance, consider making 'contributed actively to diversity and inclusion' a criteria for *all* awards. No one in the organisation should win an award or be given a reward unless they are actively part of developing and sustaining a diverse and inclusive workforce.

Building capacity for the future

The reason I am reluctant to provide a checklist for enhancing diversity, equity and inclusion across the workforce, beyond Hamilton and team's sage ideas, is that this book is about uncertainty – and acknowledging our diversity and variability is central to that. The goal is not to homogenise the workplace, but to diversify it; to add productive forms of uncertainty rather than suppress it.

In a 2021 *Forbes* article titled 'How to actually hire for diversity', an applied behavioural scientist known as an innovator in healthcare offers ways to engage a diverse workforce. While some of her recommendations are sound, I disagree with her approach. I will discuss these in the context of cultural literacy and uncertainty tolerance because I think it illustrates some important points.

The beginning of the article suggests a series of what I would label as 'uncertainty suppression' strategies. Aline Holzwarth terms these strategies ways to 'weed out bias' in the workplace. The take-home of this section is that workplaces should 'treat everyone equally'. To accomplish this, she suggests standardised checklists, standardised interview questions and objective hiring protocols. I am hopeful that by this point in the book you are already coming to a similar conclusion about these recommendations I did – that despite our best efforts and a strong desire to achieve 'objectivity' and 'standardisation', these types of strategies typically preference those who conform to the status quo. Those who are already marginalised by society can be further excluded through such approaches.

We aren't all the same. If the goal is to genuinely hire for diversity, using approaches that don't acknowledge these differences – in fact, that seek to erase them – are unlikely to get us the outcome we seek. This is because, importantly, these approaches don't eliminate unconscious bias, and can actually reinforce it. I would suggest that, for some of those in positions of power, an obsession with standardisation helps to placate feelings of discomfort that accompany discussions and initiatives around increasing diversity and inclusion in the workplace. We may tell ourselves that we can't possibly be biased because we are 'treating everyone the same' and 'levelling the playing field'. But we are failing to acknowledge the ways in which standardisation itself involves bias. We could ask who decides which questions are appropriate? How are answers standardised and according to what principles? Who is defining what 'equal treatment' means in an interview process? Who defines the 'objective hiring criteria' – is it even possible to have 'objective hiring criteria'? Have those who are marginalised by existing systems of inequity been included in these decisions?

The real work isn't in developing objectivity, but in evaluating the processes that are determining these practices. The premise of cultural hegemony is that we struggle to make objective decisions because the influence of the dominant social group is so pervasive; attempts at objectivity typically only end up reinforcing existing social patterns, with those at the top of the social hierarchy maintaining power.

Now to the good bits of the article. The author offers suggestions to correct for bias – which could, when employed thoughtfully, prove useful. Generating 'specific diversity targets', for instance, will mean that underrepresented groups are more likely to be represented than without these targets. She also suggests that hiring for multiple roles at once can be helpful, as there tends to be greater exclusion when there is only one position available. Wise.

Ultimately, hiring processes are just one way to address social inequities; the entire workplace culture needs to become supportive of diversity through inclusive practices from hiring right through to retirement. A 2019 World Economic Forum article highlights why every single aspect of the working environment needs to be reflected on and reconsidered in order to support true diversity.

Take mentorship, for instance. In 2022, Drs Gary W. Ivey, a researcher in the field of leadership and workplace culture, and Kate Dupré, an associate professor in organisational psychology, published a review on the impact of mentoring practices in workplaces. They define mentorship as someone with more experience

supporting someone with less experience. Ivey and Dupré identified two key types of support that mentors can provide their
mentees (or protégés): career-related mentorship, which focuses
on the skills, steps and processes required to succeed in a given
workplace; and psychosocial support, which enhances the mentee's professional identity, helps them find acceptance, challenges
imposter syndrome and supports their mental wellbeing. There
are also two mentorship methods: formal and informal, with formal initiated and driven by the workplace, and informal managed
by the individual, often in partnerships described as developing
'naturally' from commonalities.

The literature on mentorship suggests that, overall, mentor–
mentee relationships are invaluable. Ivey and Dupré identify
many benefits in their literature review. The positive outcomes
of effective mentorship for the mentee included: fast-tracked
upward mobility, pay rises, enhanced feelings of belonging at
work, reduced burnout and greater overall wellbeing. The authors
found that mentors also benefit from the experience by enhanced
productivity and a deep sense of satisfaction.

Not all mentor–mentee relationships are healthy, though.
Ivey and Dupré found many instances of 'the dark side of mentoring', involving 'negative relations (e.g. bullying, incivility, social
undermining), sabotage (e.g. revenge, career damage), difficulty
(e.g. conflict, ultimatums, or forced choices, such as career over
family), spoiling (e.g. betrayal, lack of fair treatment), submis-
siveness (e.g. submissive behavior by protégée in exchange for
organizational rewards), deception (e.g. manipulation by either
protégée or mentor), and harassment (e.g. sexual harassment,
sexual/racial discrimination)'. When we consider that many people from minority groups are forced to deal with abuse, bullying,

silencing and harassment on a regular basis in daily life, we should not be surprised that these dynamics can creep into workplace mentoring. Some may be denied mentoring at all. Having a formal mentoring program in a workplace, as opposed to expecting employees to rely on informal networks, is one way to help prevent unconscious bias, particularly if the diversity in a workplace is limited. Training for mentors is also essential to help avoid a toxic mentoring relationship. Ivey and Dupré provide a helpful list of things to consider when developing a workplace mentoring program, including how the mentors are selected, supported and evaluated. With a thoughtfully devised, purposeful and reflective mentorship process, the benefits that come with mentorship will be the norm.

Yet in reviewing the literature on mentorship through the lens of cultural hegemony, I have come to realise that relying solely on mentorship to support workers isn't enough. In the absence of a workplace environment that is inclusive and equitable, tolerant of social uncertainties and capable of critical reflection, mentorship could simply perpetuate the status quo. The mentors themselves may be part of the privileged group. They also likely only have experience within a certain type of workplace. Thus, their advice to their mentees could easily focus on conforming, as opposed to supporting them to develop a sense of belonging, speaking up and challenging norms.

In 2001, emeritus professor of business administration Dr David A. Thomas wrote a valuable article in *Harvard Business Review* titled 'The Truth About Mentoring Minorities: Race Matters'. In this pivotal piece, Thomas notes two critical factors leading to exodus of highly qualified diverse staff from workplaces: 'whites tend to fast-track early' and the other being mentorship.

He discusses how minorities in the workplace appear to benefit most from mentoring that includes psychosocial support. Thomas reports that 'minority executives [who had] influential mentors [who] were investing in them ... prevented them from either ratcheting down their performance or simply leaving'. When the mentoring relationship was supportive, particularly in the context of an employee who was stuck and not progressing, having someone more senior who valued and supported them seemed to counter (at least a little) other factors, including the inequitable system in which they worked. But the key revelation is that the point of difference for minority executives versus white executives and those in middle management was mentorship and support from a 'broader range of people, especially in the early years of their career'.

Thomas identified that when the mentor was white and the mentee was a person of colour, the mentor tended to focus on career-related mentorship as opposed to the needed psychosocial mentorship. This may be due, in part, to a fear to tread into a challenging discussion of race and racism in the workplace. Digging deeper into the literature, Thomas reports that white mentors' lack of psychosocial support for mentees of colour may be due to challenges in 'forming, developing and maturing' relationships. The reasons behind these challenges included negative stereotyping; fears related to connecting with those who have different experiences from our own; and negative public perceptions. There may be hesitations in building the intimacy needed for a successful mentorship, or mentors may fail to offer higher-risk but potentially rewarding opportunities in order to 'protect' the mentee from adversity. Thomas found, however, that when a white mentor could 'understand and acknowledge race as a potential barrier', their mentees did better, presumably because the mentor could engage in supporting their mentees with the complexities they faced.

Thomas's article has many outstanding recommendations for supporting those typically marginalised in workplaces. One is publicly 'endorsing the mentee's ideas' to help the mentee feel more able to take necessary workplace risks. Mentors can also offer support and recommendations that leave room for differing experiences (for example, 'This might not work for you, but from my experience …'). While having mentors who look like us and who have similar experiences to us is valuable, there appears to be merit in having diverse mentors. By engaging in cultural literacy practices and learning to navigate our uncertainty, mentors in positions of power may be able to successfully support a more diverse workforce.

What we can see is that leadership, hiring and workplace practices all need to be re-evaluated, and not just once but continually, to support diversity in the workplace. Society changes over time – and so must our systems.

The values and culture of a workplace need to move from supporting the status quo to rejoicing in a willingness to question our assumptions about people's names, languages, appearance, socioeconomic status, gender, sexuality, ethnicity, culture, religion and so on. We need to learn to leave biases at the door and, instead of valuing the person who 'knows all the answers', reward those who practise critical reflection and encourage openness and flexibility. The employee that seems the most challenging may be the exact person that should be at the table making decisions. In the hiring process, this might look like a post-interview committee discussing not only their top choices, but *why* the top choices were made, and exploring what biases might contribute to these decisions (for example, if someone is labelled as 'unprofessional', do cultural biases contribute to this view?). Focusing on the specifics of how the person could likely perform the duties of the role may also help, as refrains like 'they won't fit in' or 'they just don't seem like the right *match*' could be code for 'let's hire the same person we always do'.

How to become an anti-racist

One framework for addressing bias and inequity in the workplace is featured on the John T. Milliken Department of Medicine website. Titled 'Becoming an Anti-Racist', it involves concentric circles, labelled from inner to outer: the fear zone, the learning zone and the growth zone. Each zone features thoughts and behaviours that individuals in that zone display, showing how to move from fear to growth, and ultimately create an environment more supportive of inclusion, diversity and equity. The text accompanying the figure makes clear that personal change is necessary in order for the system to change. For example, instead of staying in the fear zone, where we 'avoid hard questions', 'seek comfort' and/or 'deny racism is a problem', we can move into the learning zone, where we recognise the problem and 'understand [our] own privilege' in being able to ignore racism, and work towards educating ourselves. While the learning zone is a step forward, we need to be in the growth zone in order to change systems. Here we 'speak out when we see racism', 'sit with our discomfort' and even 'yield positions of power to those otherwise marginalised'.

What I see in this figure, as a researcher exploring the role of uncertainty tolerance in our lives, is repeated evidence that effective management of personal uncertainty is critical to our ability to progress towards a just society. We, especially those in privileged and powerful positions, need to be aware that discomfort is part of the process. Not just that, but we need to seek it out in order to move from fear to growth.

In our workplaces, a form of regular debrief, centred on interrogating our decisions, and exploring the decision-making process around candidates, job performance and promotions, should be built into our practices. This debrief can only be effective if it is also undertaken by diverse teams.

We could also lift a page from the scientific journal community. Many journals are now asking both authors and reviewers to answer a simple question: 'How was diversity and inclusion considered in this study?' Workplaces could make this a feature of every key workplace interaction – how was diversity and inclusion considered in this meeting? In this awards ceremony? In this presentation?

At my own workplace, we have a similar question in our standardised teaching peer-review process. Among items such as 'learners are made aware of prior knowledge' and 'learning outcomes are clear' is the item 'where appropriate, Indigenous knowledges and ideas are incorporated into a lesson'. While this may be an imperfect attempt at inclusion, every single educator I have peer-reviewed over the past three years has mentioned seeing this item, and each was sparked to explore their discipline to learn about more about Indigenous knowledges in the context of what they teach. This item, embedded in our regular activities, sparked critical reflection and action to change practices.

You may have noticed that all of these suggestions start with acknowledging our own positions within the social system, particularly those of us with privilege. Stephanie Nixon, associate professor of physiotherapy at University of Toronto, provides a framework that can help us do just this. She describes this 'Coin Model' in her 2019 paper:

> There are norms, patterns and structures in society that work for or against certain groups of people, which are unrelated to their individual merit or behaviour. Put another way, there are (often invisible) systemic forces at play that privilege some social groups over others, such as sexism, heterosexism, racism, ableism, settler colonialism, and classism ... In the Coin Model, each system of inequality is conceptualized as a coin. Coins do not reflect the individual behaviour of good or bad people. Rather, they are society-level norms or structures that give advantage or disadvantage regardless of whether individuals want it or are even aware of it. Each coin represents a different system of inequality.

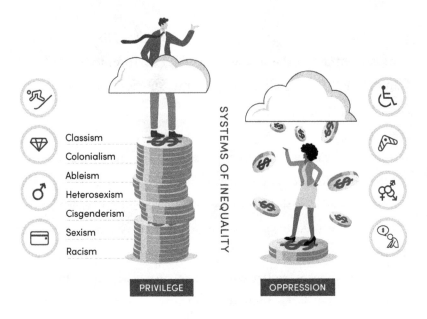

SYSTEMS OF INEQUALITY

Classism
Colonialism
Ableism
Heterosexism
Cisgenderism
Sexism
Racism

PRIVILEGE

OPPRESSION

Coins represent the systems of inequality. Those on top of the coin are in a position of privilege. Beneath these coins are those that are oppressed.

Those of us with privilege have an advantage others do not, which we did not earn and which we hold because of who we happen to be, Nixon tells us, and we stand atop a stack of coins. These privileges, in western societies, include being upper or middle class, white, a settler, able-bodied, straight, cis and male. Those who are oppressed by these systems of inequality, alternatively, have the weight of the coins bearing down upon them, as they have sources of disadvantage others do not, which they experience because of who they happen to be. In our society, this includes people who are LGTBQAI+, disabled, Indigenous, not white or from a lower socioeconomic group. The systems of inequality creating the oppression are classism, settler-colonialism, ableism, heterosexism, cisgenderism, sexism and racism, and depending on their interplay, the number of coins can vary from person to person.

If we are lucky enough to be on privileged side of the divide, we should acknowledge how these privileges affect our work. As we write a job ad, provide feedback to employees, reward staff or develop workplace culture, we can ask ourselves how our privileges may be influencing our thoughts, feelings and behaviours – and consider what we can do to counteract their effect, helping to dismantle the systems of inequality that reinforce the cultural hegemony. Through this process of critical reflection, we can begin to develop mechanisms (ladders and pulleys) to help move those oppressed to the top of larger stacks of coins. The Coin Model illustrates why standardised approaches don't work. We aren't all on the same terrain to begin with.

The Coin Model also illustrates our perspectives in society. For me, and others like me, what we see and understand about the world come from our vantage point atop many coins. When someone from across the divide speaks up and says they feel marginalised or excluded, believe them. What this person sees, what this person experiences, is different from what you see and experience. On the top of the mountain, above the rain clouds, there is only blue sky, but at the bottom of the mountain there is rain and thunder. We cannot expect someone who lives and works below the cloud cover to share our sunny vantage. This realisation may feel disconcerting and uncomfortable, calling into question our values and priorities. But do not give in to a knee-jerk reaction of intolerance to uncertainty, of inaction, of losing the opportunity for growth. Instead, listen. Believe those that report injustice, inequity and exclusion; hear them. Then do something to change it.

STRATEGIES TO SUPPORT INTERCULTURAL AWARENESS AND UNDERSTANDING

- Don't shy away from the discomfort of the uncertainty, particularly if you are in a leadership position.

- Create a system that supports uncertainty tolerance, and lean into your own uncertainty. Develop systems that question the status quo, and reward people who do this.

- Engage in critical reflection. Question yourself and ask others questions that probe the reasoning behind the structures and practices of the workplace. (For example, should I be open to running discursive, story-based interviews rather than standard question-and-answer interviews? How might our job ads be discouraging older candidates?) Use this reflective process to develop practices that support diversity in hiring and promotion.

- Create a toolkit that helps to build and sustain a workplace culture of inclusivity and diversity. (For example, an 'off-limits language' list, an inclusivity handbook, a professional mentoring program with diverse mentors).

- Make diversity and inclusion a consideration in every workplace activity, from document reviews to awards ceremonies.

- Reward those who engage in critical reflection, and create space for critical reflections in daily activities (for example, time allotted to ask critical questions during every meeting; avenues to refine the workplace's diversity, equity and inclusion strategy).

- Be aware of and sensitive to the destabilising nature of uncertainty tolerant systems and leave room for the developmental process associated with intercultural sensitivity (for example, metrics that focus on more than outputs, and include reference to time invested in supporting others).
- Regularly monitor, reassess and adjust practices and policies as the wider culture changes.
- Hold staff and peers accountable for their actions. Call out behaviour that does not support the workplace culture when you see it.
- Learn, listen and act.

Developing cultural literacy in the workplace and beyond, fostering uncertainty tolerance in others and creating inclusive practices begins with the way we teach. Teachers have the greatest power (outside of parents) to foster uncertainty tolerance in learners. Teachers can shape the way learners think and the way they approach novelty and the unfamiliar. By employing strategies that support development of cultural literacy and enhance uncertainty tolerance, we can teach learners to challenge existing social structures. We can help future workers and leaders learn to be comfortable with the discomfort that challenging the status quo brings – particularly for those in positions of privilege. Chapter 1 has more detail on relevant teaching practices to foster uncertainty, such as critical incidents, destabilisation and iso-immersion techniques.

To quote my team, the goal of cultural literacy in education is 'to help learners move away from their "bubbles" and comfort zones, as discomfort is critically important for ... enaction of social justice'. My team's findings support the existing research on cultural literacy, which suggests that it fosters uncertainty tolerance, as well as the humility and cognitive flexibility required to engage meaningfully with diversity and inclusion practices. By increasing uncertainty tolerance in workplaces and in society as a whole, we can adjust our policies and practices to be inclusive of and equitable for all, not only a privileged few.

If we accept the discomfort of uncertainty, if we create supports and spaces to actively reflect on and interrogate our biases, if we question the assumptions we make and the positions we take when faced with someone who acts or behaves 'differently' to what we know, if we embrace uncertainty and curiosity instead of rejecting them, we can create a society where everyone feels seen and valued, and fits in. Then we all win.

6

THE AGE OF UNCERTAINTY

How uncertainty tolerance helps us
navigate local and global challenges

*'Uncertainty is an uncomfortable position. But certainty
is an absurd one.'* — VOLTAIRE

W hy are so many of us afraid of uncertainty? We have
seen throughout this book how suppressing and
ignoring uncertainty can negatively impact on the
work we do, on workforce wellbeing, on cultural awareness and on
social inclusion. We have seen how uncertainty-avoidant systems
generate inequities and burnout, perpetuate injustices and biases,
and increase financial burdens while decreasing efficiency – among
other effects. While the architects of our systems and structures
crave certainty, these systems and structures are actually stronger
if we build in provisions for uncertainty.

And yet, instead of creating flexible and adaptable systems, we
continue to pursue conformity and rigidity. By trying to eliminate
uncertainty, we end up with inflexible processes that lead us along
predetermined paths to unsurprising outcomes. We perpetuate
the status quo. In contrast, freedom, creativity, opportunity and
hope are all bound to uncertainty.

Before I get too carried away, I must remember how we got here. What motivates many in their attempts to avoid and suppress uncertainty is a perception that uncertainty and risk are synonymous. The sectors in our society with greatest responsibility for human wellbeing – such as healthcare and education – seem to be particularly driven to avoid, minimise or eliminate uncertainty.

What the research suggests, however, is that most often, we can't control risk by attempting to suppress uncertainty. If we manage to suppress the identified source of uncertainty, another source stealthily appears elsewhere. We, as individuals, leaders and a society, expend copious amounts of energy attempting a losing battle of whack-a-mole. What can we do, as individuals, as leaders, as a community, to change our relationship with uncertainty? To generate adaptive, instead of maladaptive, responses to it?

Uncertainty – a penalty or a bonus?

The opening quote by Voltaire highlights the constant challenge humans face in our relationship with uncertainty. We seem to be almost programmed to avoid the unknown. But why? Does it serve us evolutionarily?

A 2022 article in *The Conversation* offers one explanation. It suggests that we have an almost innate drive to suppress uncertainty due to 'the free energy principle'. Within the thermodynamic world, 'free energy' refers to the energy left in a system that can be used towards activities that are valuable to supporting the system and its parts. The free energy principle suggests that all living things are programmed to minimise or eliminate surprises – because these types of encounters would waste energy on activities that are counterproductive to survival. The article describes these 'surprising experiences' as those 'that have not been encountered previously by the organism or its ancestors' – essentially, surprising experiences, at least in my reading of the article, appear to be synonymous with confronting uncertainty.

The idea here is that these surprising experiences are not ones that lead to the sustainability of our species (otherwise we would survive them, making them no longer a surprise or a source of uncertainty). Thus, we have evolved to minimise encounters with these surprising experiences. Our evolutionary development tells us that engaging in surprising experiences, generating uncertainty, could end in death, even the end of our species. The weight of such encounters means that we make every provision possible to suppress, ignore and avoid that which is uncertain, as individuals and as a species.

While the free energy principle is hotly debated (ironically, its own status is uncertain), it presents a valuable, albeit extreme, perspective on humans' instinctive tendency to perceive uncertainty negatively, and begins to provide a potential explanation of our almost compulsive drive to reduce, suppress or ignore uncertainty in the world around us and within ourselves.

The free energy principle may explain our innate fear of uncertainty at a species level. When considered at an individual level, the theory may take on a different aspect. Many of us seek out surprising experiences because of the certainty of anticipated uncertainty. Whether it is at the mild end of the spectrum, such as reading or watching mysteries (I am an avid watcher of all the BBC mystery television shows), or the extreme, such as skydiving; whether the thrills of a spine-tingling slasher movie or the adrenaline rush of a high-stakes poker game, these experiences are predicated on uncertainty, and yet people flock to them. Those who seek out such experiences do so in pursuit of excitement.

I am a huge Philadelphia Eagles fan (that's American football – gridiron – for those of you not in the know). I have often found a way to watch at least some of the games live here in Australia. There are some days when, due to the extreme time difference, all I learn about the game is the final score. I could watch the game after learning the score, but every single time I elect not to. *What's the point?* I think to myself. The pleasure I get from watching the game in real time is linked to the excitement of not knowing what's coming, of engaging in a shared source of uncertainty. I watch filled with hope for my team alongside other fans who are as tense as I am.

It turns out there is data to support what's going on here. Uncertainty in some situations actually prolongs our experiences of happiness, when compared against the same situations without the uncertainty. A team of researchers, led by emeritus professor of psychology Timothy D. Wilson, explored a series of studies evaluating the role of uncertainty in particular life experiences, evaluating the resulting emotions as well as our ability to predict these feelings ahead of time.

The first study explored how we perceive the uncertainty related to random acts of kindness (RAK). Participants were given RAKs by someone who was a random stranger to them, and some participants were provided certainty about the RAK while others were gifted uncertainty along with the RAK. Those receiving the 'certain' RAK were issued a gift of a silver dollar alongside very explicit information about who was providing the gift and the motivation behind the gift-giving. Those receiving the 'uncertain' RAK also received a silver dollar, but the card attached only included the name of an organisation (for example, The Smile Society), but with no other explanation or identifying information.

Another set of participants were asked to predict which of these two groups would get the most enjoyment out of the gift: those who received the gift under the certain or the uncertain conditions. Which do you think had the greatest happiness at receiving the gift, they were asked – the group where the motivation was unclear or the group who knew the details about the who, what and why behind the random act of kindness? It turns out, the group where the motivation was unclear were most surprised by the gift and experienced the greatest positive effects on mood. Contrary to the free energy principle, which suggests that humans *always* seek to avoid surprises, this study suggested that spoilers (like knowing the score of a game before watching it) tend to ruin the uplift in mood that can accompany encounters with uncertainty.

What about those predictions? Are we good at acknowledging the positive impact uncertainty can have on our future experiences?

It turns out, we aren't great at it. The majority of participants incorrectly predicted the outcome of the uncertainty, assuming that recipients would feel negatively about the gift that came with uncertainty and more positive about the gift that came with certainty.

The same researchers undertook a follow-up study exploring individuals' reactions to uncertainty when watching a movie. *Rudy* (1993) is a biopic about Daniel 'Rudy' Ruettiger – an uplifting story about a kid who beat the odds to live out his dream of being a football player. After watching the movie, participants were introduced to either certainty or uncertainty about what happened to the main character after the conclusion of the story. All participants were provided two accounts of Rudy's post-movie future, but these accounts were inconsistent. One group was told which of these accounts was true, while the other group were left to decide which was the accurate version of Rudy's future after graduation.

Their findings? While both groups felt positively about the movie, those in the uncertain group held *positive views for a longer period* following participation. The researchers also explored the behaviour of the two groups after they received the information about Rudy's fate. The uncertain group seemed to think about Rudy's future for longer periods of time than those in the certain group, suggesting that the uncertainty provoked a curiosity that led to more critical thought about the information presented.

Yet again, however, this study revealed that another set of participants were terrible at predicting the positive impacts of uncertainty on the viewers' mood. Despite these participants predicting that the conflicting views of Rudy's future would cause discomfort, the group seemed to respond positively to this uncertainty.

A 2022 study led by Professor Freda Marie Hartung, a German expert in psychological diagnostics and differential and personality psychology, explored how curiosity influences knowledge acquisition by using a series of scales that evaluate participants' epistemic curiosity, defined as a 'desire for knowledge that motivates individuals to learn new ideas, eliminate information-gaps, and solve intellectual problems', and general knowledge. While most of their participants were female, so we need to be aware of extrapolating to the universal, their results do suggest at least a correlative link between curiosity, knowledge gains and reasoning capacity.

What could this mean within the context of the previous study? Could the conflicting information about Rudy inspire the participants to reframe their relationship with the uncertainty – to take on more of an investigatory role, probing and mulling over the information, reasoning through which of the options may be true and untrue? Is this role responsible for the pleasant mood the participants experience? If so, could a reorientation towards curiosity in the face of uncertainty support both our ability to critically evaluate and learn from the uncertainty, and improve our own wellbeing?

A 2007 study titled 'Quantity versus uncertainty: when winning one prize is better than winning two' further adds to the picture, highlighting that uncertainty isn't necessarily a bad thing. Led by Jaime Kurtz, a professor of psychology at James Madison University, with contributors Professor Timothy Wilson and award-winning social psychologist Daniel Gilbert, this study explored a related experiment to those led by Wilson's team. The premise of this study included uncertainty and certainty in the receipt of gifts. Participants in the 'certain' group knew what gifts would be received and that they would receive two gifts. Those in the 'uncertain' group not only wouldn't know what they would receive, but would only get one gift. Participants were randomly assigned to one of these two groups. The study also included a group who were questioned about which outcome would produce a better mood for the participants.

The only (maybe) surprise here? That the results were the same as those in the previous studies. Participants forecasted greater satisfaction for those in the certain group, but it was those in the uncertain group who were more likely to feel better for longer periods of time due to receiving a gift. Those in the uncertain group not only didn't know what they were getting, but also received less – and still felt more positively than those who received two gifts but were aware of what they were receiving.

These studies, alongside the conclusions I draw in each chapter of this book, provide some potentially powerful insights into why our systems may be set up to suppress and minimise uncertainty. Just like the modelling we see in business and economics, humans appear to be dreadfully inept at predicting the outcomes of uncertainty, repeatedly tending to view certainty as the better of the two options – even when the uncertainty is linked to positive outcomes, like gifts and money.

When we consider that most workplaces and industry sectors seek to forecast uncertainty, to accurately predict the future of lives and assets, it appears more understandable and a little less surprising that the response is an overwhelming intolerance of uncertainty. These studies suggest that mentally rehearsing the idea of future uncertainties could result in a predisposition to negative anticipation of this uncertainty. In response, we seem to be driven to develop rigid frameworks to generate the certainty we think is better. These collective results suggest to me that we should move into uncertainty only with curiosity, without aiming to predict the outcome.

Why, then, does organisation after organisation appear to shy away from uncertainty – even when workplaces are a prime location to foster social change and enhance uncertainty tolerance at a collective level? How do we go about reorienting our relationship with uncertainty? To move from predicting negative outcomes to experiencing curiosity about the uncertainty?

Rethinking our relationship with uncertainty

A 2011 paper from the Norwegian School of Management in Oslo, Norway, led by Professor Jan Terje Karlsen, explores the ways that different sectors can support uncertainty tolerance.

In the introduction, the authors underscore a theme of this book – that effective management of uncertainty is only possible if the organisation is set up to do this. As we have seen throughout this book, an individual trying to manage uncertainties within an uncertainty-intolerant system is like a fish swimming upstream. The result is often employee exhaustion and burnout.

To avoid this, the system needs to be designed to acknowledge the uncertainty. When we are alert to uncertainty, we can identify it and factor it into our planning.

In their 2011 paper, the research team describes conducting extensive interviews that led to some eye-opening conclusions about how organisations can foster systemic and individual uncertainty tolerance. They identified the need for a reframing and reorientation towards uncertainty, moving from a suppressive and fearful mindset to a curious one, which provides opportunities to work through the unknowns. What is clear from their study is that this approach requires a wholesale transformation – uncertainty must be sought out and encouraged by the leadership, cascading down until it is embedded into the smallest tasks conducted by frontline workers. As we saw in relation to social inclusion and diversity in the workplace, those in power are the ones best placed to drive the necessary change.

The leadership, however, needs to provide employees with the support and skills to engage with uncertainty. Key moderators of uncertainty are needed to facilitate uncertainty tolerance, including a positive attitude towards uncertainty and a belief in their ability to navigate it; time and resources focused on seeing uncertainty (for example, opportunities for reflection, a dedicated time for employees to raise potential sources of uncertainty) and an openness and respect for those that ask questions and engage in critical reflection. This includes allowing the acknowledgement that sometimes 'shit happens' – as is the nature of working in a world where uncertainty is omnipresent. To quote a participant in this study, there is a need for 'a high level of tolerance for bad risks', which 'makes it easier to communicate uncertainty information'.

Just as we saw with navigating uncertainty in education, giving employees who face large amounts of uncertainty clear responsibilities and roles also appears to help build an uncertainty-tolerant organisation. The authors found that having a supportive culture is a critical moderator for improving employees' uncertainty tolerance – and this includes being able to 'recognize the existence of uncertainty as an inherent part of undertaking projects'. Interestingly, a key to building a workplace culture that is tolerant of uncertainty is to focus on the positive outcomes that can accompany uncertainty, such as opportunities to be creative and generate novel solutions.

What was most surprising to me is that there appears to be a great deal of alignment between their findings and those of my team, from methods for fostering uncertainty tolerance in the classroom to building uncertainty tolerance in workplaces and society. Many of the features of uncertainty-tolerant learning environments are mirrored in uncertainty-tolerant organisations, such as pastoral care, dedicated time to manage the uncertainty (and build up opportunities to practise uncertainty tolerance), and clear roles and responsibilities. What this reinforces for me is that we may be able to move the needle towards a more

uncertainty-tolerant society by creating opportunities to practise and embed uncertainty from primary school through to the
workplace. Key across each of these environments is the role that
individuals can play in developing an uncertainty-tolerant culture.
Given this, how can we supercharge our relationship with uncertainty in our own lives?

Living well with uncertainty

Nearly all models of uncertainty tolerance include provision
for an uncertainty stimulus, our perception of uncertainty, and
our responses to it. Some models suggest that our emotional
response drives the behavioural response, while others suggest
that we have an evolutionary drive to be less tolerant to uncertainty, which shapes our perception and responses. All of these
theories focus on the critical role that our central nervous system
(brain and spinal cord) plays in managing these uncertainties,
and the significant impact that our emotional and physiological
responses have on our actions and thoughts when uncertainty
is present.

A team of researchers led by Dr Eric Anderson, an experimental psychologist and an expert in emotional influences of
experiences, authored a paper that described a few different theories which explain the emotional and affective responses humans
seem to have to uncertainty. Many of these theories posit that
emotional and affective responses have an impact on not only
our autonomic nervous system, but also on our behaviours and
body language. Our decisions and decision-making processes
both influence our actions and are influenced by our actions.
Ultimately, most of these models suggest that our brains work
ceaselessly to make predictions. We appear to be perpetually

seeking certainty despite certainty being an illusion, as no person (or computer) can map the unknown.

One model they describe, the modal model of memory, suggests that 'emotions are generated by (1) a situation that is (2) attended to, and then (3) appraised, which creates (4) an emotional response'. Embedded in this and related theories is the idea that we adjust our emotions based on the cues we receive from the context we are in. We might appraise one situation as positive (watching a football game with friends, for example) and another as negative (walking alone in the dark, perhaps). How we appraise the context determines our behaviour. And as we learn more information, we adjust our emotional response, which in turn impacts on our behaviour.

Anderson and his team use the example of 'Jill', a woman approaching a dark alley. She might look around her for threats, her heart rate might increase, her muscles might tense. If she sees a shadow or hears footsteps in the alley, she might process this physiological response to the uncertainty stimulus as fear and flee. But what if Jill was instead approaching the dark alley to look for a secret nightclub? Then she may interpret the footsteps as someone who can help her find the door, and thus her muscles would relax as she feels relief at finding assistance, maybe even excited by the opportunity to learn the whereabouts of this new destination.

Another study, published in 2019 in the journal *Neuroscience and Biobehavioral Reviews*, conducted a meta-analysis of fMRI studies, collating the findings of multiple studies that evaluated the 'neural signatures supporting uncertainty in different contexts'. fMRI studies allow researchers to see which regions of the brain are active under different scenarios and conditions, helping them to model brain activity (neural signatures). In this meta-analysis – led by senior postdoctoral research fellow Dr Jayne Morris from the University of Reading, whose research focuses on uncertainty and emotion – the team found that 'uncertainty arising from associative learning' (or learning through experience) was most strongly linked to forebrain-associated regions, which are regions that play a role in emotion.

Interestingly, this meta-analysis also identified that our brains may be more active 'under conditions of uncertainty' than certainty, and that the parts of our brain which activate under these conditions have little overlap – reinforcing the deeper thinking observed in participants who experienced uncertainty about Rudy's future, and the reason behind the difference in mood between the two groups.

What all of these studies strongly suggest is that an emotional response to uncertainty is not only expected – it may be necessary to our management of uncertain contexts, and can drive our behaviours and potentially what we learn from an experience.

In an interview for a piece in *Discover* magazine in November 2021, Dr Anderson underscores the shared anxiety that many feel in the face of uncertainty, and suggests that distraction is a poor way to handle it. The article notes, 'Whether this is binge-watching a TV show, partying with friends or scrolling on social media, distracting yourself rarely helps you deal with the uncertainty – it just puts off the feeling of discomfort until later.' Suppressing and ignoring uncertainty or actively seeking certainty through avenues such as information-seeking can actually introduce new unknowns (as we saw with 'incidentalomas' in Chapter 3) and generate more anxiety, without addressing the uncertainty that is already there.

Instead, the article's experts advise, as I do throughout this book, to confront the uncertainty directly. In doing so, you can better plan and prepare for steps you will take to navigate it. They also suggest mulling over your emotional response – most often this is negative – and interrogating why you feel this way. Interestingly, the article mentions the same moderating factors we identified in education to help face uncertainty, such as pastoral care, a community to support us through our feelings of fear and doubt, and generating 'psychological safety'. Their final two recommendations? 'Reframing uncertainty as possibility' and finding 'meaning in the mystery' – both key moderators, we have found, that have a positive impact on a learner's uncertainty tolerance.

It is important to remember that not every field, nor every individual, can bear the same tolerance for uncertainty. In healthcare, seeking out uncertainty with abandon could be seen as not only uncaring but also dangerous. Other fields, like science, sometimes need to start from a position of certainty to be able to find the uncertainty, as the landscape is so complex that without this certain starting point we couldn't move ahead to discovery. For those marginalised in society, there may be structures that hinder or prevent the development of key moderators such as psychological safety. For many of us, though, being armed with the awareness that we can influence and change our responses to uncertainty can be both empowering and a relief. Indeed, those in positions of power can use this knowledge to build social capacity and move us collectively towards a greater tolerance of uncertainty – which may be critical, given the current existential sources of uncertainty we are facing as a species.

Navigating uncertainty in planetary health

When we consider our climate-changed future – where floods, bushfires and droughts are common, where sources of food and housing security are in question – how we confront and navigate uncertainty may change not only the way we feel but also our capacity to respond effectively and appropriately to these challenges.

Climate change, planetary health and biodiversity are looming sources of future uncertainty for our society. When I think about these environmental uncertainties and review the literature exploring related uncertainties I am struck by the enormity of it all. What we have done to the planet is grave. There are the scientific uncertainties related to the interpretations of climate-change data (what the impact of different trajectories could be); financial uncertainties as resources dwindle and change; uncertainties stemming from the impact of political decisions about planetary health (like whether we will be able to limit global warming to 1.5 degrees); and medical uncertainties about the future impact of climate change on human health (including the likelihoods for novel illnesses and pandemics). There are social uncertainties about how we will manage climate change locally, nationally and internationally, with increasing potential for many populations of climate refugees who are displaced due to rising sea levels and other forms of disaster.

Every chapter of this book, all the sources of uncertainty I identify and the strategies to navigate them, relate in some way to the overall uncertainty facing humans in the context of our climate emergency. The continuation of our species – our collective future – is the existential source of uncertainty that hangs over all else. And managing the challenges that come along with the knowledge of our current global situation is no easy task.

A team of researchers in biodiversity addressed these broad challenges in a 2014 paper published in the journal *Nature Conservation*, titled 'Towards a different attitude to uncertainty' – an apt title, given our topic. Led by acclaimed German environmental researcher Guy Pe'er, the abstract summarises the sources of uncertainty relating to climate change:

> The current rapid and ubiquitous environmental changes, as well as anticipated rates of change, pose novel conditions and complex dynamics due to the many **sources of uncertainty** are **difficult** or **even impossible to reduce**. [my bolding]

Let that sink in. The uncertainties related to biodiversity and climate change are not only numerous already, they are growing exponentially, and our chance to reduce them is all but passed. So why are we expending so much time, resources and energy on suppressing them?

I consider this article a must-read for anyone interested in the relationship between uncertainty and biodiversity. The authors describe potential negative impacts of 'seeking certainty', as well as the positive role that accepting, even embracing, uncertainty can have in this context. There are links between their findings and many of the ideas posed in this book.

As we discussed in Chapters 3 to 5, attempts to minimise uncertainty by reducing a complex system to 'simple' rules can lead to unintended consequences. Pe'er and team equate this simplification with 'overconfidence', which they define as 'incomplete knowledge as if it was absolute truth'. By reducing the varied and complex challenges to biodiversity down to mathematical equations and calculations of percentages, we convey that the solution is equally simple – when indeed solutions are complex and rife with potential uncertainties.

We saw how in healthcare the 'normal ranges' associated with bloodwork are meant to help healthcare providers determine how to manage patients. But these simplified measurements can, for some healthcare providers, be seen as absolutes instead of guides. The challenge is that when healthcare providers are given these 'norms', much of the information underpinning them is missing. Rarely do the reporting of these norms include which populations they apply to and which they don't. This can result in overconfidence and an inappropriate determination of whether someone's bloodwork is 'abnormal' or 'normal' – leading to potentially inappropriate care.

Within the field of ecology, this same type of oversimplification can result in a 'demand for simple rules of thumb', which, similar to the healthcare example, can lead to incorrect assumptions and results. The authors discuss how such 'magic numbers' lead to misleading calculations, resulting in inappropriate recommendations for fishing and hunting quotas. In some cases, these simplified equations end up supporting overfishing and decreasing marine biodiversity.

In both cases, if the features and limitations of such guidelines were transparent and the uncertainties of the calculations were put front and centre, practitioners and policymakers might be able to more accurately know when (and when not to) use these guidelines, reducing the side effect of overconfidence.

Another 'perverse effect of seeking certainty' in relation to planetary health that the authors raise is uncertainty ignorance. By 'seeking certainty at all costs', we can end up with something known as 'premature closure', which is a tendency to close down inquiry too early by focusing exclusively on a single solution once it has been identified.

In medicine, this type of bias is often linked to diagnostic errors. In conservation work, it can lead to underexplored areas of research.

The authors describe how rare ecological events ('black swans') are filled with uncertainties that modelling ignores, simply because the data is lacking. These black swans are often viewed as 'improbable and irrelevant' but can turn out to be profound triggers of critically important ecological events. By seeking certainty and a solution to climate change, we ignore the uncertainty at our peril.

The authors describe 'non-linear or complex' black-swan events, using examples such as ecological tipping points, where an ecosystem exceeds its capacity to cope with environmental shifts, resulting in a change from a steady state to a new state. If these terms (non-linear, complexity, changes in state) sound familiar, they are. I described their relevancy to uncertainty in the business and economics sectors.

Essentially, similar to the discussions of closed modelling in Chapter 3, if we ignore the black-swan events in ecological modelling, we see 'shocking' and 'unexpected' outcomes in the real world, not because we *can't* see the uncertainty, but because we *don't*. The modelling is designed to filter it out, encouraging us to prematurely close in on a solution without all of the relevant information. A potential solution to this is to combine traditional ecological modelling with complexity ecological modelling, to allow some of the unknowns and unexpecteds to be factored into the modelling – just as I and others suggest doing when addressing uncertainties in economic modelling.

Attempts to seek certainty can also prevent progress towards addressing the uncertainty. As Hillen and team described in their conceptual model, inaction is a potential result of an intolerance of uncertainty. Pe'er and team describe how the climate change debate too often seems to centre on two questions (and the same is true now, almost ten years on): '[1] Whether climate change is occurring (including speed and severity), and [2] whether it is caused or significantly facilitated by [human] factors.' By seeking certain answers to these questions *before* implementing measures to address climate change, we are setting ourselves back indefinitely – generating new uncertainties with every passing day. This is an example of inappropriate information-seeking. We don't need to know the answers to these questions to begin tackling climate change.

Martin Péron, a data scientist, explores this logic in his 2018 PhD thesis, 'Optimal sequential decision-making under uncertainty'. Péron suggests that the process known as adaptive management ('learning by doing') may be an appropriate way to manage decision-making in the face of this level of unprecedented uncertainty. He writes that climate-change policy could be developed, deployed as we observe the outcomes, and adjusted as necessary. A similar approach is used in classrooms through a process known as 'action research', where a teaching intervention is deployed, data about its effectiveness is collected in real time, and adjustments are made as the intervention continues, not afterwards.

A group of researchers further expanded on this approach in their 2011 research review, 'Decision-making under great uncertainty: environmental management in an era of global change'. Led by Professor Stephen Polasky, an expert in biodiversity conservation and climate change from the University of Minnesota, the review similarly underscores the unprecedented breadth of uncertainty related to climate change due to the 'rate of change', a lack of historical data to rely on for future predictions and the complex systems involved. It asks the question, 'How can one best guide decision-making to meet present and future human needs in an era of global change?'

Professor Polasky and his team highlight the limitations of traditional decision theory in this context, as it relies heavily on data that is unlikely to be available, as well as on the capability to narrow down probabilities of future states, which is essentially impossible given the level of unprecedented uncertainty related to climate change. The authors recommend adaptive management, which treats 'decisions ... as experiments, often involving active participation by interested parties, that generate information that can improve future decisions'.

This seems to me to be very similar to healthcare contexts, where sometimes the sheer number of uncertainties related to a patient are too large to predict the final outcome with absolute certainty. Uncertainty tolerance in this context might be defined as being able to take the next best step with the information available, and learning as the patient progresses. If we think of climate change as a health emergency and the planet as the patient, the similarities begin to become even clearer. Just as in healthcare, you have a threshold where a patient's health is likely to be at risk if an intervention is not made – and once you cross that threshold, it can be too late. Certainly, the danger becomes that much worse.

Polasky and team suggest that a thresholds approach to decision-making could prove useful in climate change by 'focusing attention on critical boundaries that have major consequences if crossed'. The focus on reducing emissions by imposing targets is an example of this. There is enough data – more than enough data – to suggest that allowing continual CO_2 emissions to go unchecked is likely to have disastrous impacts on planetary health that we are negligent if we do not act on this knowledge.

In complexity scenario planning, we assume that when we run scenarios of our future in a climate-changed world, we will get a different outcome with every computation, resulting in as many types of potential interventions. Here, the data put into the modelling can include scope for black-swan events, rather than excluding them. Threshold theory allows us to act today and complexity modelling allows us to better prepare for tomorrow – a powerful combination in the face of the uncertainty that risks our planet and our species.

The last piece of the puzzle when it comes to making decisions under uncertain conditions is what the authors refer to as 'resilience thinking'. This final step relies on diverse perspectives to help see and manage the uncertainty, but at its core it really summarises what this book is about: we need to build, as the authors state, 'capacity to adapt to changing conditions' to 'recognize and respond to emerging transformations as they occur'. The authors suggest that leaders should look for the 'weak' early signals that the unexpected is coming, and act on this. Again, we see how those in charge are the ones best placed to see and lead a response to the uncertainty – in this case, to save our planet.

This also means fostering an environment where the hiccups aren't quietened. Similar to a culture of safety in healthcare, a culture of climate safety will encourage us to raise these early signals.

This points to something central: the power of individuals. Individual responses to uncertainty are what collectively lead to change and support of the health of the planet. Pe'er and his team argue that viewing uncertainty as risk, as so many in decision-making roles seem to do, can lead to more harm than good. The authors describe how farmers in the past would rely on local knowledge and community approaches to preventing floods. Many of these mitigation strategies had 'positive ecological and social' side effects. Now, however, farmers can turn to an insurer to help in managing their financial risk if an unexpected event occurs, and this has indirectly 'replaced the necessity for ecosystem-based buffers'. Just as we saw in the healthcare sector, when we conflate uncertainty with risk, we may be indirectly working against ourselves. Here, some farmers are moving away from the very actions necessary to address climate-change fallout and increasing the risk to the health of the planet – understandably, because insurance mitigates the huge financial loss when unexpected things go wrong, and there isn't the same certainty in these ecological approaches. Of course, many have insurance but engage consciously in land-management strategies that work to minimise the environmental footprint of farming and even restore the soil to health. Pe'er and team describe some of the positive impacts of uncertainty tolerance on improving planetary health. Yet again, these mirror the benefits described in sectors throughout this book. As we saw in Chapter 4, the very nature of scientific discovery is predicated on uncertainty. By embracing uncertainty in exploring sustainable practices and interventions around climate change, we generate pathways to develop solutions, as well as hope that solutions are still possible. Both of these things are necessary in the treatment plan for planetary crisis.

Tomorrow isn't certain and that's okay – post-normal science

In an age of 'post-' (postmodernism, poststructuralism, post-truth) there is yet another 'post' to add to the list: 'post-normal science'. In their research, Robert A. Francis, professor of urban ecology and society at King's College London, and Michael Goodman, a professor of environment and human geography, explore the relationship between the numerous sources of uncertainty stemming from climate change, including what this means for the science and how to understand and address it. In their 2010 paper in *Journal for Nature Conservation*, they argue for the value of a contemporary form of research known as post-normal science.

This concept was introduced towards the end of the last millennium, in 1993, by distinguished philosophers of science Silvio Funtowicz and Jerome Ravetz. Its premise is that our traditional western approaches to science research (for example, the reductionist approaches described in Chapter 4) are unable to address urgent contemporary social challenges, where the known knowns are far and few between and the unknowns are substantial.

In Chapter 4, I described how traditional western science research starts with an unknown, but then, through the steps of the scientific method, uncertainties are sequentially reduced and the risks related to the research are minimised as best as possible. Professors Francis and Goodman describe how climate-change research is hampered by this traditional approach to scientific inquiry. In the context of climate change, the unknown unknowns are so great that these rigid (almost dogmatic) scientific approaches, which are dependent on some certainty, become irrelevant. Instead, they argue, 'post-normal science' is the way forward to address the expansive uncertainties – too numerous to reduce, suppress or eliminate, and whose solutions cannot occur in the 'absence of risk'. Think of a terminally ill patient. Many support interventions at this stage have a high degree of risk attached to them, given the greater risk of death. Is our planet at the terminal stage? If not, when will that tipping point occur?

Post-normal science argues that complex problems require diverse inputs to develop effective solutions. Professor Ravetz discusses the definitional characteristics of post-normal science and compares these to traditional western science models in his 2004 paper published in the journal *Futures*. In this paper, Ravetz briefly describes the problematic nature of traditional western approaches to science, including that the sources of research funds are often external to the scientific community, the 'research community' is elitist and there are considerable ethical concerns around monetising scientific findings through patents and such, when many of these scientific so-called 'discoveries' are actually developed from traditional knowledges, and involve the participants of the research projects themselves. Is it ethical for the scientist to get a patent, and the resulting financial gains, when the findings leading to that patent would have been impossible without the participants?

Ravetz argues that in contemporary society 'we can no longer separate "nature", "science" and "society"'. Climate change and its impacts require coordination of these three elements. So the traditional approaches of western scientific research are unlikely to be useful because 'the reductionist tradition of western science [assumes] that complex systems are … capable of being taken apart, studied in their elements and then reassembled'. In the face of system-wide and ever-increasing uncertainties, in the context of a complex system, it will get us nowhere. Post-normal science, on the other hand, starts from a point of the unknown and uncertain, and values debate about solutions. Instead of seeking to suppress uncertainty, post-normal science seeks to invite it in.

When is the right time to use post-normal science? Ravetz states, 'when the 'stakes [are] high and [the] decisions are urgent'. Based on this, post-normal science seems the right tool at the right time for the job of identifying a potential treatment plan for the planetary health emergency.

Post-normal science is an approach to research that is co-designed with those inside and outside the formal scientific community. Unlike traditional western science, where those trained in the field conduct and communicate the science, in post-normal science those outside of the scientific community – citizen scientists – are actually needed to address the scientific questions. Francis and Goodman suggest that an 'extended peer community' is critical to the contemporary scientific challenge we face with climate change. They explain why:

> From this definition, the science of conserving species and ecosystems is certainly post-normal; despite the abundance of rigorous scientific investigation that is performed by the global scientific community, cost of the decisions taken that relate to conservation policy, funding and implementation are made by non-scientists while the physical implementation of conservation practices (usually at a local scale) is often performed by individuals with limited (or outdated) scientific training, often on a voluntary basis.

During the Covid-19 pandemic, we saw how an extended peer community can change the course of human health. It wasn't enough to rely on scientists and public health experts. We needed the expertise of community leaders to bridge the gap between what we were learning through the science and what government policies were being made as a result of that science, with what individual citizens engaged and embraced in their communities.

Dr Holly Seale, a social scientist and expert in infectious diseases and public health, led a team to explore the role of community leaders and 'other information intermediaries' during the pandemic. Published in the journal *Humanities and Social Sciences Communication*, the study underscored how failure to consult with community leaders in planning and in communicating solutions meant that the implemented policies often failed to 'resonate with the target communities'. Their work illustrates the value of post-normal scientific approaches in addressing contemporary, urgent sources of social uncertainty.

Who can or should be part of this extended community necessary for effective post-normal science practices? Ravetz suggests anyone and everyone who wants to contribute: interested individuals, those with 'local knowledges', even 'investigative journalists'. These diverse perspectives can be combined with published literature, but the research isn't predicated only on traditional sources of data. Ultimately, a key tenet of post-normal science approaches is, as Ravetz states, 'a process of mutual learning among those with different perspectives and commitments, including scientists themselves'. The extended peer community isn't there to be talked at, but rather to be engaged from the start in defining the research problem, investigating outcomes and implementing solutions.

I grew up listening to my dad tell stories about people from all walks of life having a significant impact on the trajectory of one project or another. One such story was about a large chemical manufacturing plant located very far north in the United States. Every winter the temperatures would plummet and ice would form on the powerlines, preventing essential electricity getting to the factory. Each year, despite knowing the inevitability of this outcome, the company could never figure out how to prevent the powerlines from going down from the ice build-up on the wires. It was costing the company millions, so my dad would say.

Finally, the company had had enough. They decided to get a team together to solve the puzzle of the freezing powerlines. The problem-solving team wasn't just made up of electrical engineers – anyone who worked at the company was engaged in the discussion, from the cleaners to the COO (chief operating officer). After running through very expensive ideas, or ideas that were impractical because of lack of resources or complicated designs, a friend of my dad's, a technical support person in one of the labs, wondered out loud, 'How could we get birds to fly over the site and shake the ice off the powerlines?'

This very simple question, posed by a non-expert, started an entirely new line of inquiry and thinking. The discussions moved away from high-tech, expensive and time-consuming solutions to ones that were lower in cost and immediately available, because a non-expert asked a really good question.

What was the solution they settled on that kept their business open and running for decades to come? A helicopter that flew above the powerlines to loosen the ice.

When I listened to these stories, I thought they had a sprinkle of the tall tale. Thanks to Google, though, I have learned that some helicopter companies actually advertise this as a service! Now, I understand that the company appeared to engage in post-normal science before it was even a thing, engaging an extended peer community to solve a problem that had plagued them for years.

Ravetz emphasises that the role of an extended peer community in decision-making 'is not a matter of [forming] a committee consensus' but 'a process of common discovery of creative solutions to complex situations where "compromise" becomes not so much a surrender of a prior advantage as a sharing of benefits from an arrangement of mutual aid'. Post-normal science aims to dismantle ethnocentrism in the scientific community, in that it removes the view that only those who are formally part of the scientific community can offer value to research. It flips the traditional script where the scientific community is the 'in-group' and those in the broader community are the 'out-group' – in post-normal science, anyone engaged in the research process is part of the in-group. My dad's helicopter story taught me this. A complex, ongoing problem was solved through a diversity of ideas and shared exploratory processes.

What post-normal science does do, however, is introduce potential for more sources of uncertainty. As we add more humans to the research process, we potentially introduce more uncertainty, both due to the individual freedom of human choice, and due to the uncertainties related to interpersonal interactions and communications. In a cyclical manner, post-normal science and adaptive decision-making processes all require, at some level, an individual capacity to effectively manage uncertainty – that is, uncertainty tolerance.

How to build our uncertainty tolerance

The findings from my team's own research, alongside the broader literature, can offer some support to helping us to build our own uncertainty tolerance. Increasing our uncertainty tolerance may not only improve our own wellbeing; it may even save our species. As we have seen throughout this book, our own approaches to managing uncertainty can influence our workplace culture, our peer group and broader society.

What we want to avoid is what I call 'uncertainty overload', where the uncertainty we experience is so constant that we allow ourselves to become overwhelmed and paralysed by the extent of the uncertainty. When we enter this zone, we become stuck, stagnating, steeped in the uncertainty, unable to identify solutions and next steps. In the context of the pandemic, this might manifest as denial – a refusal to take the impacts of the virus seriously – or, alternatively, as an inability to make decisions about the future. For climate change, this may look like not taking steps to change our behaviour or approach to business until 'enough information is in'. What may be occurring during these phases of uncertainty overload is that we focus on seeking a single, all-encompassing solution and think too far into the future. Instead, we can learn from adaptive decision-making, where we focus on taking the next step, acknowledging that this step isn't the end point, or even perhaps the move determining our fate – we can always turn around and try a different path. By knowing our own thresholds, by understanding the point at which we tip into uncertainty overload, we can identify when is the best time to take the first step in responding to uncertainty – and reassure ourselves that whatever step we take is not fatalistic. When we take a considered first step, and are open to adjusting the path if the outcome isn't helpful, we can incorporate adaptive decision-making processes into our daily lives.

A first step is to reframe uncertainty as a potential for personal growth and an exercise in choice. It is easier to take that first step when we aren't picturing a negative result, but rather curiosity about the future. There are many other strategies we can use that draw on my team's research, as well as on the other research presented throughout this book, as we have seen. I've used many of these practices to help myself, and encouraged my staff to do the same, to manage situations of widespread uncertainty.

A simple yet powerful step we can take to manage uncertainty effectively is to **find our purpose**, no matter how small, to light our path through the ambiguity – or as Anderson puts it, to 'find meaning in the mystery'. Clarifying your role and your position in managing the uncertainty can help. Teachers and employers can both help to engage this moderator by outlining clear roles and responsibilities for their students and employees.

Critical reflection is the one moderator that seems to appear across every study as a powerful influence. In the context of our own journeys through uncertainty, this can mean identifying opportunities and knowledge that come from navigating uncertainty, and defining how we will apply these to future uncertainties we will (not might) encounter. This can also mean reflecting on cultural norms and identifying the foundations of these norms in society's decision-making processes. Ongoing critical reflection can help us engage in uncertainty-seeking behaviours that, in turn, may allow us to be better prepared to address uncertainty and can help make the unexpected a little more anticipated. Workplaces can support this type of activity by building in dedicated time to practise critical reflection, and having the leadership team purposively seek critically challenging questions and feedback from their employees.

What we can learn from my team's research in education, and the literature on decision-making in the face of uncertainty, is that problem solving to navigate unknowns and uncertainties appears to be best guided through diverse teamwork. While a simple manta of 'embrace diversity' isn't the answer, a conscious and collective effort to create practices that **foster inclusion** of diverse knowledges, opinions, perspectives and approaches appears to be an important contributor in addressing wide-ranging uncertainties. In classrooms and workplaces, we can use demographic information to generate diverse learner and employee teams to tackle complex and challenging questions. The premise of post-normal science depends on these extended and diverse communities to help both identify the uncertainties and work to solve them.

Communication of uncertainty is another key factor to building uncertainty tolerance. My team found that during the pandemic, learners felt they could manage uncertainty more effectively when they were kept in the loop frequently, even when the communication conveyed that the answers and directions were still uncertain. Knowing that their teachers were aware of the unknowns and were exploring potential solutions gave them enough certainty to navigate the uncertainty. We see similar outcomes in healthcare in communication with patients. These studies seem to illustrate that few patients expect certainty in their healthcare, but they are better able to manage uncertainty when their healthcare provider highlights where the uncertainty exists and what the next step is in their care plan. In science research we see that, similarly, communication of uncertainty can help society better understand the research findings.

Creating a supportive environment is essential. The psychological studies exploring uncertainty highlight the strong (often negative) emotional response we tend to have when faced with uncertainty, particularly when this uncertainty could pose a threat to ourselves or our community. To manage this, fostering a sense of belonging, of psychological safety, is key. When we connect at times of personal vulnerability, it not only helps us to reduce that vulnerability but it encourages us to form strong connections that will endure.

What my team's research reveals, and other research seems to reveal indirectly, is that moderators of our uncertainty tolerance do not work in isolation. Instead, we should use multiple moderators together. This will help us to generate a community, a society and a planet that is better equipped to face the constant stream of uncertainties that will define our future, and to better address the urgent uncertainties facing us today, including the climate emergency. Ultimately, we all benefit when individuals are better prepared to manage uncertainty effectively, when systems are equipped to balance uncertainty with risk, and when education is designed to help us build skills necessary to achieve this at a young age.

A few months ago, a tweet came up in my Twitter feed that stuck with me. In the aftermath of a terrible personal grief, one user had written: 'Life is fragile, we live in a chaotic and random universe. Life is a gift of unknown duration.' We need to treat each other with kindness because ultimately the only certainty in our lives is uncertainty, is the point he was making. He was right.

The world we live in is uncertain. If we accept this, rather than pretending it is untrue, we can ready ourselves for the shocks, and position ourselves to experience the joy. We can't predict the future, but we can learn to navigate it. And that is the true power of the uncertainty effect.

ACKNOWLEDGEMENTS

The research and writing of this book took place on the unceded lands of the Boonwurrung/Bunurong and Wurundjeri peoples of the Kulin nation. I pay my respect to their Elders past and present, and extend this respect to all Aboriginal and Torres Strait Islander peoples, including those who gave of their time and knowledge (either through their own writings or by electing to participate in the research) to help us all learn and develop.

As I discuss in Chapter 4, one person's work bears the marks of many others. This book is no exception. Its words are influenced by all those I have encountered – even some I have yet to meet IRL – who have inspired me through their writing, tweets and commentaries. This work is based on my own and others' lived experiences, as well as on primary research studies, and so I am indebted to countless individuals. Despite the number of sentences I have written in this book, I am finding that words have limited capacity to convey the gratitude I feel to so many. I will do my best.

First, I would like to thank my partner, Matt. I know that my ability to fulfil my workaphile (*not* workaholic) lifestyle is dependent on my greatest supporter, champion and love. To him, I am eternally grateful, and I will forever love him three.

The pandemic challenged many of us as we sought to connect with others through isolation, lockdowns and border closures. I was not immune to the effects of such restrictions. The distance between me and my family in the United States seemed infinitely larger during 2020 and the following years. This book ended up creating a bridge of sorts. My dad grew up in a generation where dyslexia was a label that hampered educational and professional prospects. Despite his brilliance, a higher degree didn't pan out for him, partly because of the educational challenges he faced and partly because he had to work full time to help support the family. Upon entering the workforce, my dad did all sorts of jobs, from manual labour to food services, military service (through the draft) and administrative roles. His breadth of real-world experience, and his tendency towards lateral thinking, made him an ideal first reviewer for each chapter of this book. He never held back when giving feedback and seemed to always have a poignant story to tell, many of which are woven throughout this book. I am eternally grateful for this opportunity to connect with my dad during a time of such disconnection. I will forever remember with love and affection the early-morning debates and discussions we had as we unpacked complex ideas. He challenged me to think differently and to write clearly – and so I did.

I am also eternally grateful to the entire team of colleagues who took the time to review sections of this book, provide expert insights and championed key concepts, including Professor Amanda Berry, Dr Bob Cvetkovic, Dr Georgina Stephens, Dr Mel Farlie, Lisa Bryant, Associate Professor Francine Marques, Dr John Clarke, Associate Professor Simon Angus, Dr Jaai Parasnis, Professor John Bertram, Dr Mandy Truong, Dr Kristen Noble, Dr Amanuel Elias and Dr Gabriel García Ochoa. The selfless hours that each of these individuals put in, and the words of encouragement as I drafted and re-drafted chapters, were the light I needed in what sometimes felt like a dark tunnel.

My team's own research findings are woven throughout this book. This research is a direct result of participants who volunteered their time, insights and words of encouragement. Many have reached out to express how just being part of the research has changed the way they see and interact with the world – which contributed to my motivation to write the book. Without them this book quite literally could not exist. Thank you, each and every one of you. You inspire me, and our team, to keep going.

The premise of this book, and my team's body of research into uncertainty, was sparked by the students I teach. We had to face a stark reality, after analysing data early on, that our teaching approach was inadequately illustrating to our students the importance of human variation (that is, uncertainty). As with so many teaching innovations, wanting to do better by our students became our motivation for change. My team's research focused on how to better prepare our passionate and brilliant students for the uncertainty they will experience in their future careers. The lessons we learned were profound and led to real change, and I remain eternally grateful for the students who continue to inspire us all.

This research was possible due to the dedication of an entire research team, and dedicated funders (including Medical Education and the Monash University Education Academy), who remain committed to helping progress knowledge. It really does take a village. My village are talented, insightful, thoughtful, creative and wonderful humans, including some mentioned above who reviewed and/or contributed quotes to this book, and those that contributed to the research projects cited within the book, including Dr Amany Gouda-Vossos; Associate Professor Gabrielle Brand; Dr Angela Ziebell; Associate Professor Elizabeth Davis; Dr Swati Mujumdar; Professor Tina Brock; Professor Paul White; Dr Mahbub Sarkar; Dr Chantal Hoppe; Dr Nazmul Karim; Dr Adam Wilson; and Dr Shemona Rozario.

I also want to thank my editor, Monash University Publishing director and publisher Julia Carlomagno. Having only written academic books and journal articles up to this point, I had no idea of the intellect, influence, insight, information and introspection (this alliteration is all for you, Julia! Edit away) an editor brings to the process, and the effect it can have on a writer. I am eternally grateful for Julia's time, vision and intuition. Her expertise and knowledge of the field, and her willingness to work with my naivety, alongside her guidance and belief in this project, are what kept me motivated during the periods of frustration and writer's block. Thank you – this book exists because of you.

There are also those people who influence and impact you personally, that see in you something you don't always see, that bring the smiles, passion and energy to help you continue on, and that inspire you through their support and role modelling. Thank you to everyone who vocally and quietly supported this work – you know who are, and I can never express in words how much I appreciate you.

BIBLIOGRAPHY

Introduction

Hillen, Marij A., et al. 'Tolerance of uncertainty: conceptual analysis, integrative model, and implications for healthcare.' *Social Science & Medicine* 180 (2017): 62–75.

Chapter 1

Land, Ray and Meyer, Jan H.F. 'Threshold concepts and troublesome knowledge: linkages to ways of thinking and practising within the disciplines.' *Improving Student Learning: Improving Student Learning Theory and Practice–Ten Years On.* Oxford Centre for Staff and Learning Development (2003).

Meyer, Jan H.F. and Land, Ray. 'Threshold concepts and troublesome knowledge (2): epistemological considerations and a conceptual framework for teaching and learning.' *Higher Education* 49.3 (2005): 373–88.

Lazarus, Michelle D.; Gouda-Vossos, Amany; Ziebell, Angela; and Brand, Gabrielle, et al. 'Fostering uncertainty tolerance in anatomy education: lessons learned from how humanities, arts and social science (HASS) educators develop learners' uncertainty tolerance.' *Anatomical Sciences Education* 16 (2022): 128–47.

Lazarus, Michelle D.; Truong, Mandy; Douglas, Peter; Selwyn, Neil. 'Artificial intelligence and clinical anatomical education: promises and perils.' *Anatomical Sciences Education* (2022): 1–14.

García Ochoa, Gabriel and McDonald, Sarah. 'Destabilisation and cultural literacy.' *Intercultural Education* 30.4 (2019): 351–67.

Stephens, Georgina C.; Rees, Charlotte E.; and Lazarus, Michelle D. 'Exploring the impact of education on preclinical medical students' tolerance of uncertainty: a qualitative longitudinal study.' *Advances in Health Sciences Education* 26.1 (2021): 53–77.

Stephens, Georgina C.; Sarkar, Mahbub; and Lazarus, Michelle D. 'Medical student experiences of uncertainty tolerance moderators: a longitudinal qualitative study.' *Frontiers in Medicine* (2022): 985.

Lazarus, Michelle D. *Preparing today's learners for uncertainty.* myfuture Insights series. Education Services Australia (2021).

Ange, Brittany L., et al. 'Pass/fail grading in medical school and impact on residency placement.' *Journal of Contemporary Medical Education* 9.2 (2019): 41.

Reed, D.A., et al. 'Relationship of pass/fail grading and curriculum structure with well-being among preclinical medical students: a multi-institutional study.' *Academic Medicine* 86.11 (2011): 1367–73.

Jordan, Michelle E.; Kleinsasser, Robert C.; and Mary F. Roe. 'Cautionary tales: teaching, accountability, and assessment.' *The Educational Forum* 78.3 (2014): 323–7.

Tai, Joanna; Boud, David; and Bearman, Margaret. 'Grading students may be as easy as ABC, but evidence shows better ways to improve learning.' *The Conversation*, 31 March 2022.

Precel, Nicole. 'Swinburne dumps grades, says marking hinders creative process.' *The Age*, 4 March 2022.

McClain, Lauren; Gulbis, Angelika; and Hays, Donald. 'Honesty on student evaluations of teaching: effectiveness, purpose, and timing matter!' *Assessment & Evaluation in Higher Education* 43.3 (2018): 369–85.

Helsing, Deborah. 'Regarding uncertainty in teachers and teaching.' *Teaching and Teacher Education* 23.8 (2007): 1317–33.

Floden, Robert E. and Buchmann, Margret. 'Between routines and anarchy: preparing teachers for uncertainty.' *Oxford Review of Education* 19.3 (1993): 373–82.

Mackay, Margaret and Tymon, Alex. 'Working with uncertainty to support the teaching of critical reflection.' *Teaching in Higher Education* 18.6 (2013): 643–55.

Chapter 2

Pomare, Chiara, et al. 'A revised model of uncertainty in complex healthcare settings: a scoping review.' *Journal of Evaluation in Clinical Practice* 25.2 (2019): 176–82.

Eachempati, Prashanti, et al. 'Developing an integrated multilevel model of uncertainty in health care: a qualitative systematic review and thematic synthesis.' *BMJ Global Health* 7.5 (2022), e008113.

Strout, Tania D., et al. 'Tolerance of uncertainty: a systematic review of health and healthcare-related outcomes.' *Patient Education and Counseling* 101.9 (2018): 1518–37.

Han, Paul K.J., et al. 'How physicians manage medical uncertainty: a qualitative study and conceptual taxonomy.' *Medical Decision Making* 41.3 (2021): 275–91.

Ho, Emma Kwan-Yee, et al. 'Psychological interventions for chronic, non-specific low back pain: systematic review with network meta-analysis.' *BMJ* 376 (2022).

Slade, Susan C.; Molloy, Elizabeth; and Keating, Jennifer L. 'The dilemma of diagnostic uncertainty when treating people with chronic low back pain: a qualitative study.' *Clinical Rehabilitation* 26.6 (2012): 558–69.

Han, Paul K.J., et al. 'Uncertainty in health care: towards a more systematic program of research.' *Patient Education and Counseling* 102.10 (2019): 1756–66.

Hancock, Jason and Mattick, Karen. 'Tolerance of ambiguity and psychological well-being in medical training: a systematic review.' *Medical Education* 54.2 (2020): 125–37.

Levi, Ryan and Gorenstein, Dan. 'How a clinician's desire to be thorough can cause a harmful spiral of unnecessary care', WVPE, 10 June 2022.

Medendorp, Niki M., et al. 'A scoping review of practice recommendations for clinicians' communication of uncertainty.' *Health Expectations* 24.4 (2021): 1025–43.

Lazarus, Michelle D, et al. 'The human element: how educators can prepare learners for future workplace uncertainties and troublesome knowledge.' In Davies et al. Koninklijke Brill NV. Expected publication 2023.

Truong, Mandy, et al. 'Resisting and unlearning dehumanising language in nursing and healthcare practice, education and research: a call to action.' *Nurse Education Today* 116 (2022): 105458.

Obermeyer, Ziad, et al. 'Dissecting racial bias in an algorithm used to manage the health of populations.' *Science* 366.6464 (2019): 447–53.

Christensen, Donna M.; Manley, Jim; and Resendez, Jason. 'Medical algorithms are failing communities of color.' *Health Affairs*, 9 September 2021.

Igoe, Katherine J. 'Algorithmic bias in healthcare exacerbates social inequities – how to prevent it.' Harvard School of Public Health, 12 March 2021.

Thomas, Angela; Krevat, Seth; and Ratwani, Raj. 'Policy changes to address racial/ethnic inequities in patient safety.' *Health Affairs*, 25 February 2022.

Burrowes, Kelly. 'Small step for Nature, giant leap across the gender gap: leading journal will make sex and gender reporting mandatory in research.' *The Conversation*, 27 May 2022.

Epstein, Ronald M. and Privitera, Michael R. 'Doing something about physician burnout.' *The Lancet* 388.10057 (2016): 2216–7.

Lewin, Evelyn. '"I can't do this anymore": how burnout affects doctors and patients.' *NewsGP*, 5 March 2021.

Galvin, Gabby. 'Nearly 1 in 5 healthcare workers have quit their jobs during the pandemic.' *Morning Consult*, 4 October 2021.

Clarke, J.R. 'How a system for reporting medical errors can and cannot improve patient safety.' *American Journal of Surgery*, 72.11 (2006): 1088–91; discussion 1126–48.

Feldpush, B. 'Implementing standardised colors for patient alert wristbands.' *Quality Advisory: American Hospital Association*, 4 September 2008.

Medline. 'ID wrist bands: color-coded for patient identification.' *Medline Industries, Inc.* www.medline.com/media/catalog/Docs/MKT/ LIT475_BRO_MedlineIDBands_horizontal.pdf.

Stephens, Georgina C., et al. 'Reliability of uncertainty tolerance scales implemented among physicians and medical students: a systematic review and meta-analysis.' *Academic Medicine* 97.9 (2022): 1413–22.

Stephens, Georgina C.; Sarkar, Mahbub; and Lazarus, Michelle D. '"A whole lot of uncertainty": a qualitative study exploring clinical medical students' experiences of uncertainty stimuli.' *Medical Education* 56.7 (2022): 736–46.

Rees, Eliot L., et al. 'Evidence regarding the utility of multiple mini-interview (MMI) for selection to undergraduate health programs: a BEME systematic review: BEME Guide No. 37.' *Medical Teacher* 38.5 (2016): 443–55.

Pau, Allan, et al. 'The multiple mini-interview (MMI) for student selection in health professions training – a systematic review.' *Medical Teacher* 35.12 (2013): 1027–41.

Truu, Maani. 'The really old, racist and non-medical origins of the BMI.' *ABC News*, 2 January 2022.

Chapter 3

Mishra, Vivek. 'RBA interest rate hike now likely in Q3, to end year at 0.50%.' *Reuters*, 25 February 2022.

Kelly, Jack. 'You'll be surprised to learn what the workers who quit their jobs in the "Great Resignation" are doing now.' *Forbes*, 15 October 2021.

Thompson, Derek. 'The Great Resignation is accelerating.' *The Atlantic*, 15 October 2021.

Serafeim, George. 'If Greece embraces uncertainty, innovation will follow.' *Harvard Business Review*, 13 March 2015.

Gupta, Parveen P. and Fogarty, Timothy J. 'Governmental auditors and their tolerance for ambiguity: an examination of the effects of a psychological variable.' *The Journal of Government Financial Management* 42.3 (1993): 25.

Makkawi, Bilal A. and Abdolmohammadi, Mohammad J. 'Determinants of the timing of interim and majority of audit work.' *International Journal of Auditing* 8.2 (2004): 139–51.

Makkawi, Bilal A. and Rutledge, Robert W. 'Evaluating audit risk: the effects of tolerance-for-ambiguity, industry characteristics, and experience.' *Advances in Accounting Behavioral Research (Advances in Accounting Behavioural Research, Vol. 3)*. Emerald Publishing (2000): 69–89.

Dulaney, Michael. 'How the "big four" accounting firms could threaten the global economy.' The Money, *ABC News*, 12 July 2018.

Wright, Michael E. and Davidson, Ronald A. 'The effect of auditor attestation and tolerance for ambiguity on commercial lending decisions.' *Auditing: A Journal of Practice & Theory* 19.2 (2000): 67–81.

Katsaros, Kleanthis K.; Tsirikas, Athanasios N.; and Nicolaidis, Christos S. 'Managers' workplace attitudes, tolerance of ambiguity and firm performance: the case of Greek banking industry.' *Management Research Review* 37.5 (2014): 442–65.

Wagener, Stephanie; Gorgievski, Marjan and Rijsdijk, Serge. 'Businessman or host? Individual differences between entrepreneurs and small business owners in the hospitality industry.' *The Service Industries Journal* 30.9 (2010): 1513–27.

Arthur, William B. 'Foundations of complexity economics.' *Nature Reviews Physics* 3 (2021): 136–45.

Schasfoort, Joeri. 'Complexity Economics.' *Exploring Economics* (2017) www.exploring-economics.org/en/orientation/complexity-economics/.

Jones, Charles I. 'The facts of economic growth.' *Handbook of Macroeconomics, vol. 2.* Elsevier (2016): 3–69.

Boeing, Geoff. 'Visual analysis of nonlinear dynamical systems: chaos, fractals, self-similarity and the limits of prediction.' *Systems* 4.4 (2016): 37.

Pavie, Xavier and Carthy, Daphne. 'Leveraging uncertainty: a practical approach to the integration of responsible innovation through design thinking.' *Procedia – Social and Behavioral Sciences* 213 (2015): 1040–49.

Banning, Kevin C. 'The effect of the case method on tolerance for ambiguity.' *Journal of Management Education* 27.5 (2003): 556–67.

Farmer, J. Doyne, et al. 'A third wave in the economics of climate change.' *Environmental and Resource Economics* 62.2 (2015): 329–57.

Sahakian, Barbara J. 'Humans vs AI: here's who's better at making money in financial markets.' *The Conversation*, 2 February 2022.

Chapter 4

Vellar, Ivo D. and Hugh, Thomas B. 'Howard Florey, Alexander Fleming and the fairy tale of penicillin.' *The Medical Journal of Australia* 177.1 (2002): 52–3.

Arseculeratne, S. N. and Arseculeratne, G. 'A re-appraisal of the conventional history of antibiosis and penicillin.' *Mycoses* 60.5 (2017): 343–47.

Pray, L. 'Discovery of DNA structure and function: Watson and Crick.' *Nature Education* 1.1 (2008): 100.

Ross, Matthew B.; Glennon, Britta M. and Murciano-Goroff, Raviv, et al. 'Women are credited less in science than men.' *Nature* 608 (2022): 135–45.

Yang, Yang, et al. 'Gender-diverse teams produce more novel and higher-impact scientific ideas.' *Proceedings of the National Academy of Sciences* 119.36 (2022), e2200841119.

Brewster, Carisa D., 'How the woman who found a leprosy treatment was almost lost to history.' *National Geographic*, 1 March 2018.

Nobles, Melissa, et al. 'Science must overcome its racist legacy: *Nature's* guest editors speak.' *Nature* 606.7913 (2022): 225–27.

Sense About Science. *Making Sense of Uncertainty: Why Uncertainty is Part of Science*. Sense About Science (2013).

Philipp-Muller, Aviva; Petty, Richard and Lee, Spike W. S. 'Understanding why people reject science could lead to solutions for rebuilding trust.' *The Conversation*, 15 July 2022.

Davis, Josh. 'There are more male than female specimens in natural history collections.' *Natural History Museum*, 23 October 2019.

Kale, Alex; Kay, Matthew and Hullman, Jessica. 'Decision-making under uncertainty in research synthesis: designing for the garden of forking paths.' *Proceedings of the 2019 CHI conference on human factors in computing systems* (2019).

Clegg, Joshua W. 'Uncertainty as a fundamental scientific value.' *Integrative Psychological and Behavioral Science* 44.3 (2010): 245–51.

Djulbegovic, Benjamin. 'Articulating and responding to uncertainties in clinical research.' *Journal of Medicine and Philosophy* 32.2 (2007): 79–98.
———. 'Acknowledgment of uncertainty: a fundamental means to ensure scientific and ethical validity in clinical research.' *Current Oncology Reports* 3.5 (2001): 389–95.

Van Der Bles, Anne Marthe, et al. 'The effects of communicating uncertainty on public trust in facts and numbers.' *Proceedings of the National Academy of Sciences* 117.14 (2020): 7672–83.

Steffensen, Victor. *Fire Country: How Indigenous Fire Management Could Help Save Australia*. Hardie Grant Explore (2020).

Pascoe, Bruce. *Dark Emu*. Magabala Books (2014).

Woolston, Chris and Osório, Joana. 'When English is not your mother tongue.' *Nature* 570.7760 (2019): 265–67.

Abbasi, Alireza and Jaafari, Ali. 'Research impact and scholars' geographical diversity.' *Journal of Informetrics* 7.3 (2013): 683–92.

Amarante, Veronica and Zurbrigg, Julieta. 'The marginalization of southern researchers in development.' *World Development Perspectives* 26 (2022): 100428.

Editorial. 'Support Europe's bold vision for responsible research assessment.' *Nature*, 27 July 2022.

Kim, Eunji, et al. 'Navigating "insider" and "outsider" status as researchers conducting field experiments.' *PS: Political Science & Politics* (2022): 1–5.

Cech, Erin A. 'The intersectional privilege of white able-bodied heterosexual men in STEM.' *Science Advances* 8.24 (2022): eabo1558.

Wu, Lingfei; Wang, Dashun; and Evans, James A. 'Large teams develop and small teams disrupt science and technology.' *Nature* 566 (2019): 378–82.

Peters, Hans Peter and Dunwoody, Sharon. 'Scientific uncertainty in media content: introduction to this special issue.' *Public Understanding of Science* 25.8 (2016): 893–908.

Wong, Sissy. 'We have to change how we teach science for the future energy workforce.' *Forbes*, 13 June 2019.

Chapter 5

Rose, Damon. 'The s-word.' *BBC News*, 12 April 2006.

Gawn, Lauren. 'Why the words matter.' Ramp Up, *ABC News*, 13 September 2011.

Miller, Barbara. 'Times dictate name change for Spastic Centre.' *ABC News*, 8 February 2011.

Zimmer, Benjamin. 'A brief history of "spaz".' *Language Log*, 13 April 2006.

Frenkel-Brunswik, Else. 'Tolerance toward ambiguity as a personality variable.' *American Psychologist* 3.268 (1948): 385–401.

MacKay, Jenna. Profile of Bonnie Strickland. *Psychology's Feminist Voices Digital Archive* (2010): 261–66.

Freidenreich, Harriet. *Else Frenkel-Brunswik.* Jewish Women's Archive (1999).

Vertovec, Steven. 'Super-diversity and its implications.' *Ethnic and Racial Studies* 30.6 (2007): 1024–54.

————. *Superdiversity: Migration and Social Complexity.* Taylor & Francis (2023): 251.

Bennett, Milton J. and Bennett, Milton J. 'Intercultural sensitivity.' *Principles of Training and Development* 25.21 (1993): 185–206.

Bennett, Milton J. and Hammer, Mitchell. 'A developmental model of intercultural sensitivity.' *The International Encyclopedia of Intercultural Communication* 1.10 (2017).

Cromb, Natalie. 'Australia is racist – but not in the way you think.' NITV, 10 August 2017.

Faa, Marian. 'Brisbane club Hey Chica! bans 23yo Moale James with face tattoos from entering.' *ABC Pacific*, 30 June 2022.

Cargile, Aaron Castelán and Bolkan, San. 'Mitigating inter-and intra-group ethnocentrism: comparing the effects of culture knowledge, exposure, and uncertainty intolerance.' *International Journal of Intercultural Relations* 37.3 (2013): 345–53.

Gramsci, Antonio. *Selections from the Prison Notebooks.* Lawrence and Wishart (1971).

Baudrillard, Jean. *Simulacra and Simulation (The Body, In Theory: Histories of Cultural Materialism)*. University of Michigan Press (1994).

Trioli, Virginia. 'No, women of a "certain age" don't need your flowers or performative charity, so take your assumptions elsewhere.' *ABC Radio Melbourne*, 16 July 2022.

Forrest, James and Dunn, Kevin. '"Core" culture hegemony and multiculturalism: perceptions of the privileged position of Australians with British backgrounds.' *Ethnicities* 6.2 (2006): 203–30.

Strong, Rowan. 'An antipodean establishment: institutional Anglicanism in Australia, 1788–c. 1934.' *Journal of Anglican Studies* 1.1 (2003): 61–90.

Murray-Atfield, Yara. 'Melbourne's Shrine of Remembrance cancels rainbow light plan, citing threats and abuse.' *ABC News*, 30 July 2022.

Hogg, Michael A. and Adelman, Janice. 'Uncertainty–identity theory: extreme groups, radical behavior, and authoritarian leadership.' *Journal of Social Issues* 69.3 (2013): 436–54.

Hogg, Michael A. 'Uncertainty–identity theory.' *Advances in Experimental Social Psychology* 39 (2007): 69–126.

———. 'Uncertain self in a changing world: a foundation for radicalisation, populism, and autocratic leadership.' *European Review of Social Psychology* 32.2 (2021): 235–68.

———. 'The search for social identity leads to "us" versus "them".' *Scientific American*, 1 September 2019.

Grant, Fiona and Hogg, Michael A. 'Self-uncertainty, social identity prominence and group identification.' *Journal of Experimental Social Psychology* 48.2 (2012): 538–42.

Wagoner, Joseph A. and Hogg, Michael A. 'Uncertainty–identity theory.' *Encyclopedia of Personality and Individual Differences* (2017): 1–8.

Smallen, Dave. 'Experiences of meaningful connection in the first weeks of the COVID-19 pandemic.' *Journal of Social and Personal Relationships* 38.10 (2021): 2886–905.

———. 'Feeling connected enhances mental and physical health – here are 4 research-backed ways to find moments of connection with loved ones and strangers.' *The Conversation*, 26 July 2022.

Elias, Amanuel; Mansouri, Fethi; and Paradies, Yin. *Racism in Australia Today*. Palgrave Macmillan (2021).

Elias, Amanuel and Paradies, Yin. 'The costs of institutional racism and its ethical implications for healthcare.' *Journal of Bioethical Inquiry* 18.1 (2021): 45–58.

VicHealth. *Counting the Billion-Dollar Cost of Racism in Australia*. VicHealth (2016).

Nyuon, Nyadol. 'Peter Dutton is the "uncivil" one.' *The Saturday Paper*, 23 July 2022.

García Ochoa, Gabriel; McDonald, Sarah; and Monk, Nicholas. 'Embedding cultural literacy in higher education: a new approach.' *Intercultural Education* 27.6 (2016): 546–59.

García Ochoa, Gabriel and McDonald, Sarah. *Cultural Literacy and Empathy in Education Practice.* Springer (2020).

Lustig, Myron W. *Amongus & Intercultural Competence.* Pearson (2012).

Lazarus, Michelle D.; García Ochoa, Gabriel; Truong, Mandy; and Brand, Gabrielle. 'Advancing social justice capacity in health professions learners through interdisciplinary education approaches: engaging uncertainty tolerance & cultural literacy.' In Brady, Jennifer and Gingras, Jennifer. *Teaching Social Justice in Health Professions Education.* University of Regina Press (expected publication 2023).

Eswaran, Vijay. 'The business case for diversity in the workplace is now overwhelming.' *World Economic Forum,* 29 April 2019.

Lorenzo, Rocío; Voigt, Nicole; Tsusaka, Miki; Krentz, Matt; and Abouzahr, Katie. 'How diverse leadership teams boost innovation.' *BCG,* 23 January 2018.

Deloitte Global. *Striving for Balance, Advocating for Change: The Deloitte Global 2022 Gen Z & Millenial Survey.* Deloitte (2022).

Weber Shandwick. *Millennials at Work: Perspectives on Diversity & Inclusion.* Weber Shandwick (2016).

Hamilton, Odessa S.; Kohler, Lindsay; Cox, Elle B.; and Lordan, Grace. 'How to make your organization's language more inclusive.' *Harvard Business Review,* 18 March 2022.

Burn, Ian, et al. 'Older workers need not apply? Ageist language in job ads and age discrimination in hiring. No. w26552.' *National Bureau of Economic Research* (2019).

Boroditsky, Lena. 'How language shapes thought.' *Scientific American,* 1 February 2011.

Holzwarth, Aline. 'How to actually hire for diversity.' *Forbes,* 18 February 2018.

Page, Marcia. 'Why getting workplace diversity right isn't for the faint-hearted.' *World Economic Forum,* 8 March 2019.

Ivey, Gary W. and Dupré, Kathryn E. 'Workplace mentorship: a critical review.' *Journal of Career Development* 49.3 (2022): 714–29.

Thomas, David A. 'The truth about mentoring minorities. Race matters.' *Harvard Business Review* 79.4 (2001): 98–107.

Tolliver, McKinzie. *Becoming Anti-Racist.* Washington University School of Medicine in St. Louis (2020).

McCluney, Courtney L., et al. 'To be, or not to be … Black: the effects of racial codeswitching on perceived professionalism in the workplace.' *Journal of Experimental Social Psychology* 97 (2021): 104199.

Nixon, Stephanie A. 'The coin model of privilege and critical allyship: implications for health.' *BMC Public Health* 19.1 (2019): 1–13.

Chapter 6

Pain, Ross; Kirchhoff, Michael David; and Mann, Stephen Francis, '"Life hates surprises": can an ambitious theory unify biology, neuroscience and psychology?' *The Conversation*, 15 August 2022.

Wilson, Timothy D., et al. 'The pleasures of uncertainty: prolonging positive moods in ways people do not anticipate.' *Journal of Personality and Social Psychology* 88.1 (2005): 5.

Bar-Anan, Yoav; Wilson, Timothy D.; and Gilbert, Daniel T. 'The feeling of uncertainty intensifies affective reactions.' *Emotion* 9.1 (2009): 123.

Hartung, Freda-Marie, et al. 'Being snoopy and smart: the relationship between curiosity, fluid intelligence, and knowledge.' *Journal of Individual Differences* 43.4 (2022): 194–205.

Kurtz, Jaime L.; Wilson, Timothy D.; and Gilbert, Daniel T. 'Quantity versus uncertainty: when winning one prize is better than winning two.' *Journal of Experimental Social Psychology* 43.6 (2007): 979–85.

Karlsen, Jan Terje. 'Supportive culture for efficient project uncertainty management.' *International Journal of Managing Projects in Business* 4.2 (2011): 240–56.

Anderson, Eric C., et al. 'The relationship between uncertainty and affect.' *Frontiers in Psychology* 10 (2019): 2504.

Morriss, Jayne; Gell, Martin; and van Reekum, Carien M. 'The uncertain brain: a co-ordinate based meta-analysis of the neural signatures supporting uncertainty during different contexts.' *Neuroscience & Biobehavioral Reviews* 96 (2019): 241–49.

Morriss, Jayne, et al. 'Intolerance of uncertainty is associated with heightened responding in the prefrontal cortex during cue-signalled uncertainty of threat.' *Cognitive, Affective & Behavioral Neuroscience* 22.1 (2022): 88–98.

Baerg, Lindsay and Bruchmann, Kathryn. 'COVID-19 information overload: intolerance of uncertainty moderates the relationship between frequency of internet searching and fear of COVID-19.' *Acta Psychologica* 224 (2022): 103534.

Sweeney, Morgan. 'Uncertainty is uncomfortable. Here's how we can learn to live with it.' *Discover*, 8 November 2021.

Pe'er, Guy, et al. 'Towards a different attitude to uncertainty.' *Nature Conservation* 8.67 (2014): 95–114.

Péron, Martin Brice. *Optimal sequential decision-making under uncertainty.* Diss. Queensland University of Technology (2018).

Polasky, Stephen, et al. 'Decision-making under great uncertainty: environmental management in an era of global change.' *Trends in Ecology & Evolution* 26.8 (2011): 398–404.

Francis, Robert A. and Goodman, Michael K. 'Post-normal science and the art of nature conservation.' *Journal for Nature Conservation* 18.2 (2010): 89–105.

Ravetz, Jerry. 'The post-normal science of precaution.' *Futures* 36.3 (2004): 347–57.

Seale, Holly, et al. 'The role of community leaders and other information intermediaries during the COVID-19 pandemic: insights from the multicultural sector in Australia.' *Humanities and Social Sciences Communications* 9.1 (2022): 1–7.